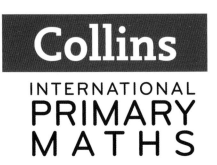

Collins

INTERNATIONAL PRIMARY MATHS

Student's Book 3

William Collins' dream of knowledge for all began with the publication of his first book in 1819. A self-educated mill worker, he not only enriched millions of lives, but also founded a flourishing publishing house. Today, staying true to this spirit, Collins books are packed with inspiration, innovation and practical expertise. They place you at the centre of a world of possibility and give you exactly what you need to explore it.

Collins. Freedom to teach.

Published by Collins
An imprint of HarperCollins*Publishers*
The News Building
1 London Bridge Street
London
SE1 9GF

HarperCollins*Publishers*
Macken House,
39/40 Mayor Street Upper,
Dublin 1,
D01 C9W8, Ireland

> **Browse the complete Collins catalogue at
> www.collins.co.uk**

10 9 8 7 6

ISBN 978-0-00-836941-5

British Library Cataloguing-in-Publication Data
A catalogue record for this publication is available from the British Library.

Author: Caroline Clissold
Series editor: Peter Clarke
Publisher: Elaine Higgleton
Product developer: Holly Woolnough
Project manager: Mike Harman (Life Lines Editorial Services)
Development editor: Joan Miller
Copyeditor: Tanya Solomons
Proofreader: Catherine Dakin
Cover designer: Gordon MacGilp
Cover illustrator: Ann Paganuzzi
Typesetter: Ken Vail Graphic Design
Illustrators: Ann Paganuzzi and QBS Learning
Production controller: Lyndsey Rogers
Printed and bound in India by Replika Press Pvt. Ltd.

With thanks to the following teachers and schools for reviewing materials in development: Antara Banerjee, Calcutta International School; Hawar International School; Melissa Brobst, International School of Budapest; Rafaella Alexandrou, Pascal Primary Lefkosia; Maria Biglikoudi, Georgia Keravnou, Sotiria Leonidou and Niki Tzorzis, Pascal Primary School Lemessos; Taman Rama Intercultural School, Bali.

The publishers gratefully acknowledge the permission granted to reproduce the copyright material in this book. Every effort has been made to trace copyright holders and to obtain their permission for the use of copyright material. The publishers will gladly receive any information enabling them to rectify any error or omission at the first opportunity.

Contents

Number

Geometry and Measure

Statistics and probability

How to use this book

This book is used towards the start of a lesson when your teacher is explaining the mathematical ideas to the class.

Key words

- The **key words** to use during the lesson are given. It's important that you understand the meaning of each of these words.

- An **objective** explains what you should know, or be able to do, by the end of the lesson.

Let's learn

This section of the Student's Book page **teaches** you the main mathematical ideas of the lesson. It might include pictures or diagrams to help you **learn**.

 An activity that involves thinking and working mathematically.

1 An activity or question to discuss and complete in pairs.

Guided practice
Guided practice helps you to answer the questions in the Workbook. Your teacher will talk you through this question so that you can work independently with confidence on the Workbook pages.

HINT

Use the page in the Student's Book to help you answer the questions on the Workbook pages.

1 **Thinking and Working Mathematically** (TWM) involves thinking about the mathematics you are doing to gain a deeper understanding of the idea, and to make connections with other ideas. The TWM star at the back of this book describes the 8 ways of working that make up TWM. It also gives you some sentence stems to help you to talk with others, challenge ideas and explain your reasoning.

At the back of the book

Number

Lesson 1: **Counting**

- Count on and count back in steps of 10 and 100

Let's learn

265

200   60   5

What happens when we keep adding hundreds?

What happens when we keep adding tens?

What pattern do you notice when you keep adding 100?
How many digits change each time?
Which digits change?
Which digits remain the same?

What pattern do you notice when you keep adding 10?
How many digits change each time?
Which digits change?
Which digits remain the same?

Guided practice

Count forwards in 10s from 362.

362 | 372 | 382 | 392 | 402 | 412 | 422 | 432 | 442

Describe the pattern.

The 1s digit stays the same. The 10s digit increases by 1 each time. After 392, the 100s digit increases by 1, and the 10s digit is 0. The pattern continues with the 10s digit increasing by 1 each time.

Lesson 2: **Even and odd numbers**

Key words
* even number
* odd number

• Recognise even and odd numbers

Let's learn

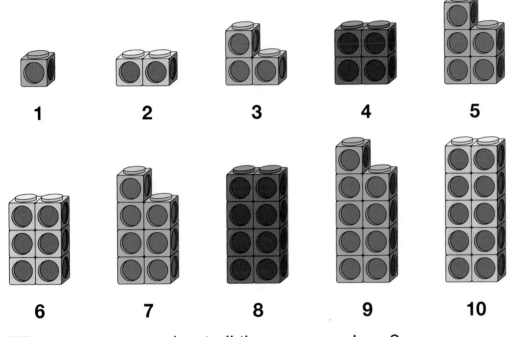

1   2   3   4   5

6   7   8   9   10

What can you say about all the even numbers?

What can you say about all the odd numbers?

What digits do even numbers end with?

What digits do odd numbers end with?

Guided practice

Sort these numbers into a Carroll diagram.

567   106   354   902   24   231   788   350   479   843   15   189

Even numbers	Not even numbers
24 106 354	15 231 189
788 350 902	567 479 843

Lesson 3: **More about even and odd numbers**

Key words
• even number
• odd number
• shared equally
• divided by 2

Number

• Recognise even and odd numbers

Let's learn

What do you notice?

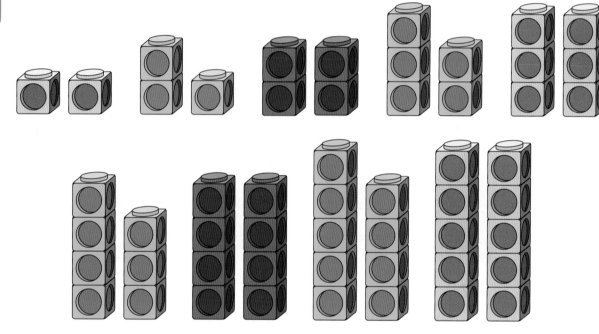

👥 What can we say about even numbers?

◁2 What can we say about odd numbers?

Is this sometimes, always or never true?

Guided practice

Draw a ring around the numbers that can be shared equally between 2.

Are the numbers even or odd? Show this in the table.

63　(24)　37　91　(72)　(86)　(138)　(50)　145　(12)

Odd	Even
145　37	24　50
91　63	72　86
	138　12

Lesson 4: **Estimating**

• Estimate the number of objects

Key words
• **estimate**
• **approximately**
• **range**

Number

Let's learn

Approximately how many jelly beans are there?
What might be a good range to estimate
how many jelly beans there are?

Maths in everyday life is
not always about exact
numbers or the exact
answer to a calculation.
It may be based on
estimates.

Talk with your partner about when you make and use estimates.
How many different examples can you think of? Write them down.

Guided practice

a Estimate how many strawberries there are.

Give your answer as a range. | 50–60 |

b Draw a ring around a group of 10 strawberries. Then keep making
groups of 10 to help you count them.

c How many strawberries are there altogether? | 56 |

Lesson 1: **More about estimating**

* Estimate quantities

Key words
* estimate
* range

Let's learn

Making estimates can be useful, for example when we need to know how much money we have.

When we do a calculation, making an estimate helps us see if our answer is about right.

There are between 40 and 60 cubes here. The range is 40–60.

If we count them one by one, we might count some cubes twice or even miss some out.

By arranging the cubes in groups of 10 we can see that there are 5 towers of 10 and 6 left over, making a total of 56.

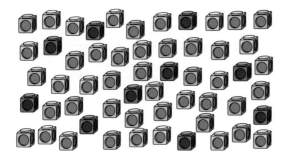

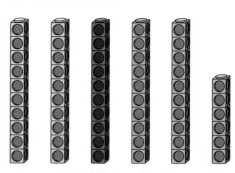

Discuss how many flowers you think there are. Give your estimate as a range. Why is this a good range to estimate the number of flowers?

Guided practice

Write a sensible range for each number.

a 48 | 40–50 | a 204 | 200–210 | a 687 | 680–690

Lesson 2: **Counting on and back**

• Recognise patterns when counting in steps of different sizes

Key words
• counting in steps
• even
• odd

Number

Let's learn

When we count in steps, we can often see patterns.

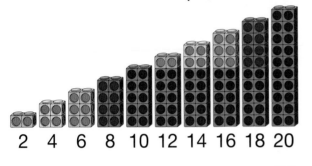

2 4 6 8 10 12 14 16 18 20

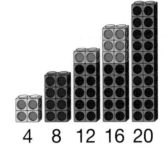

4 8 12 16 20

What do you notice about counting in 2s and 4s?

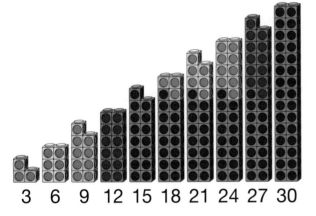

3 6 9 12 15 18 21 24 27 30

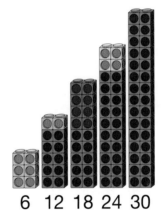

6 12 18 24 30

What do you notice about counting in 3s and 6s?

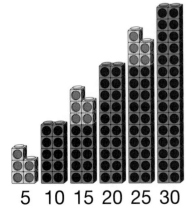

5 10 15 20 25 30

What do you notice about counting in steps of 5?
What about steps of 10?

Count in steps of 4 from zero. Why won't you say 15?
What steps could you count in to say 15?

Guided practice

Meera counts on in 5s from 30. Colour the numbers she says.

(40) (50) (54) (65) (75)

Lesson 3: **Making sequences with numbers**

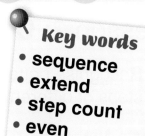

Key words
- sequence
- extend
- step count
- even
- odd

- Recognise and describe number sequences

Let's learn

Look at the sequence formed by these sweets.

2,   4,   6,   8

2 + 2   4 + 2   6 + 2

When we count in steps of 2, the rule is: Add 2 each time.

We can use number sequences with measures, such as money.
This sequence counts back by 5 cents each time.

25,   20,   15,   10,   5

25 – 5   20 – 5   15 – 5   10 – 5

What sequence do these 10 cent coins make?

Extend the sequence to the 10th group of 10 cent coins.

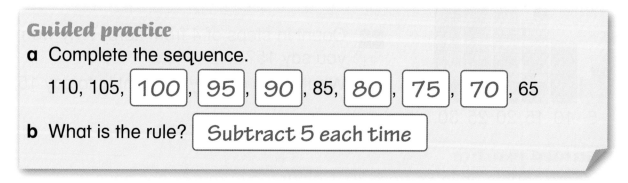

Guided practice

a Complete the sequence.

110, 105, 100 , 95 , 90 , 85, 80 , 75 , 70 , 65

b What is the rule? Subtract 5 each time

Lesson 4: **Making patterns with numbers**

• Identify patterns in numbers

Key words
• pattern
• increase
• decrease

Number

Let's learn

    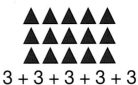

3 triangles   3 + 3   3 + 3 + 3   3 + 3 + 3 + 3   3 + 3 + 3 + 3 + 3
              6 triangles   9 triangles   12 triangles   15 triangles

The rule is 'add 3'.

Can you work out what the next two numbers will be in the pattern?

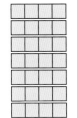  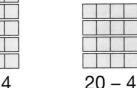

28 squares   28 − 4   24 − 4   20 − 4   16 − 4
              24 squares   20 squares   16 squares   12 squares

4 squares are taken away each time.
The rule is 'subtract 4'.

Write down the first 10 numbers you say when counting in 3s from 0.
Draw a pattern to help you.
Write down the first 10 numbers that you say when counting in 6s from 0.
Draw a pattern to help you.
Which numbers appear in both your counts?

Guided practice

a Draw the pattern:
10, 8, 6, 4, 2

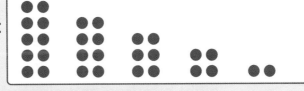

b What is the rule? Subtract 2

13

Number

Lesson 1: **Numerals and words (A)**

Key words
• **numeral**
• **number**
• **digit**
• **place holder**
• **zero**

• Read and write numbers as numerals and in words

Let's learn

764

seven hundred and sixty-four

There are 7 hundreds (700), 6 tens (60) and 4 ones (4).

439

four hundred and thirty-nine

There are 4 hundreds (400), 3 tens (30) and 9 ones (9).

It is a 3-digit number and it is even.

Is a 3-digit number and it is odd.

Use these words to make up 3-digit numbers.

| one | five | four | hundred | and | thirty |

How many different numbers can you make?

Guided practice

Draw lines to match each numeral with its written number.

503 — three hundred and sixty-seven
894 — five hundred and three
367 — seven hundred and eighty-nine
789 — eight hundred and ninety-four

Number

Lesson 2: **Numerals and words (B)**

- Read and write numbers as numerals and in words

Let's learn

246   two hundred and forty-six

(100) (100) (10) (10) (10) (10) (1) (1) (1) (1) (1) (1)

In this number there are 2 hundreds, 4 tens and 6 ones.

Take turns to use counters to make a 3-digit number.

Tell your partner what your number is and how it is made up.

Your partner then uses the numeral cards to make your number.

Both of you then use the vocabulary cards to make your number.

Repeat, taking turns to make other 3-digit numbers.

You will need
- place value counters or coloured counters in three colours (per pair)
- numbers in numerals and words cards (per pair)

Guided practice

Draw counters to represent the number 435.

(100) (100) (100) (100) (10) (10) (10) (1) (1) (1) (1) (1)

Write the number in words.

> four hundred and thirty-five

15

Number

Lesson 3: **Reading and writing even numbers to 1000**

* Read and write even numbers as numerals and in words

Let's learn

What numbers do these three sets of counters represent?

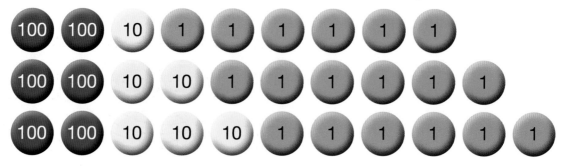

What is the same about these numbers?

What is different?

What is the sequence?

What are the next three numbers in the sequence?

What number does this set of counters represent?

If we increase the number by 2 ones, what will the next number be?

If these two numbers form part of a sequence, what are the next three numbers in the sequence?

Write the numbers as numerals and in words.

Guided practice

Write the even number in words.

678   six hundred and seventy-eight

Write the even number as a numeral.

four hundred and twelve   412

Lesson 4: **Reading and writing odd numbers to 1000**

Key words
• sequence
• odd number
• increase

Number

• Read and write odd numbers as numerals and in words

Let's learn

What numbers do these three sets of counters represent?

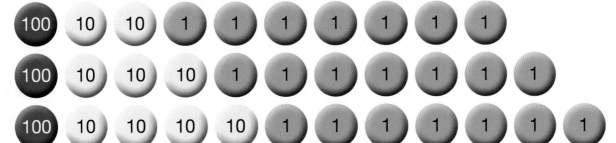

What is the same about these numbers?

What is different?

What is the sequence?

What are the next three numbers in the sequence?

What number does this set of counters represent?

If we increase the number by 2 ones, what will the next number be?

If these two numbers form part of a sequence, what are the next three numbers in the sequence?

Write the numbers as numerals and in words.

Guided practice

Write the odd number in words.

325 | *three hundred and twenty-five*

Write the odd number as a numeral.

six hundred and seven | 607

Lesson 1: **Commutativity**

- Understand that addition can be done in any order, but subtraction cannot

Key words
- commutative
- add
- sum
- total
- subtract
- difference

Let's learn

We can add numbers in any order, the sum will always be the same.

$$30 + 20 = 50 \qquad\qquad 20 + 30 = 50$$

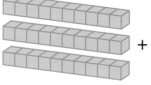 + 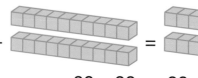 =

$$30 + 20 = 20 + 30$$

Subtraction is the inverse (or opposite) of addition.

Can we subtract in any order?

$$50 - 30 = 20 \qquad\qquad \text{Think about } 30 - 50 =$$

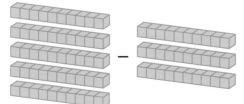 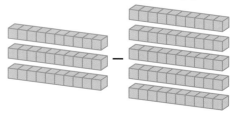

There are not enough tens in 30 to take 50 away.

We can't do subtraction in any order.

For each pair of additions, discuss which addition you think is easier. Explain why one addition is easier than the other.

| 5 + 34 | 20 + 70 | 50 + 26 | 42 + 23 | 3 + 4 + 6 |
| 34 + 5 | 70 + 20 | 26 + 50 | 23 + 42 | 6 + 4 + 3 |

Guided practice

1. Write two examples to show that addition can be done in any order.

$$40 + 20 = 20 + 40 = 60 \qquad 70 + 10 = 10 + 70 = 80$$

2. Write two examples to show that subtraction cannot be done in any order.

$50 - 20$ is not the same as $20 - 50$

$70 - 50$ is not the same as $50 - 70$

Lesson 2: **Complements of 100 and multiples of 100**

Key words
- commutative
- inverse
- add
- sum
- total
- subtract
- difference

- Know pairs of numbers that total 100
- Add and subtract multiples of 100

Number

Let's learn

If we know number pairs to 10, we know number pairs to 100 and to 1000.

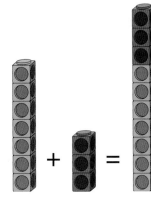

$7 + 3 = 10$ $3 + 7 = 10$
$10 - 3 = 7$ $10 - 7 = 3$

$70 + 30 = 100$ $30 + 70 = 100$
$100 - 30 = 70$ $100 - 70 = 30$

$700 + 300 = 1000$ $300 + 700 = 1000$
$1000 - 300 = 700$ $1000 - 700 = 300$

We can use known addition and subtraction facts to help us work out the answers to unknown facts.

$25 + 75 = 100$ $75 + 25 = 100$ $67 + 33 = 100$ $33 + 67 = 100$
$100 - 75 = 25$ $100 - 25 = 75$ $100 - 33 = 67$ $100 - 67 = 33$

$500 + 200 = 700$ $200 + 500 = 700$
$700 - 200 = 500$ $700 - 500 = 200$

$\square + \bigcirc = 100$ $\square + \bigcirc = 600$

Talk to your partner about which numbers could replace the square and circle. Write several facts for each. Also write each related subtraction fact.

Guided practice
Write the two addition and subtraction facts for 15 and 85.

$15 + 85 = 100$ $85 + 15 = 100$

$100 - 85 = 15$ $100 - 15 = 85$

Lesson 3: **Addition and subtraction of 2-digit numbers**

Key words
• estimate
• add
• sum
• total
• subtract
• difference

• Estimating, adding and subtracting 2-digit numbers

Let's learn

36 + 28 =

Partition 36 and 28.

Add the ones.

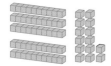

Regroup the 14 ones into 1 ten and 4 ones.

Add the tens.

$$\begin{array}{r} 3\ 6 \\ +\ 2\ 8 \\ \hline 1\ 4 \\ 5\ 0 \\ \hline 6\ 4 \end{array} \qquad \begin{array}{r} 3\ 6 \\ +\ 2\ 8 \\ \hline 6\ 4 \\ {}^1 \end{array}$$

Recombine the tens and the ones to make 64.

64 − 35 =

Partition 64 into 50 and 14.

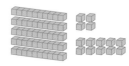

Subtract the ones.

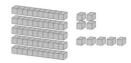

Subtract the tens.

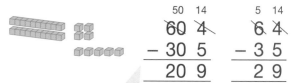

$$\begin{array}{r} {}^{50\ \ 14} \\ \cancel{6}\cancel{0}\ \cancel{4} \\ -\ 3\ 0\ \ 5 \\ \hline 2\ 0\ \ 9 \end{array} \qquad \begin{array}{r} {}^{5\ \ 14} \\ \cancel{6}\ \cancel{4} \\ -\ 3\ 5 \\ \hline 2\ 9 \end{array}$$

Recombine the tens and the ones to make 29.

Guided practice

Estimate the answer to each calculation.

a 65 + 28 Estimate: *60 + 30 = 90*

b 82 − 64 Estimate: *80 − 60 = 20*

Lesson 4: **Unknowns!**

- Use objects to represent unknown quantities

Number

Let's learn

We can use any object to represent an unknown.

 + $12 = $20

We don't know the price of the ball – it is **unknown**.

But we know that the price of the ball + $12 = $20.

We can use the relationship between addition and subtraction to work out the price of the ball.

So, if $20 – $12 = $8, then the ball must cost $8.
Let's check:

$8 + $12 = $20

$20 – = £15

We don't know the price of the doll – it is **unknown**.
But we know that $20 – price of the doll = $15

We also know that $20 – $15 = price of the doll.
So, if $20 – $15 = $5, then the doll must cost $5.
Let's check:

$20 – $5 = £15

 + $4 = $16

What is the cost of one toy car?

Guided practice
Write the unknown number in each calculation.

a $17 - \boxed{8} = 9$

b $\boxed{30} + 40 = 70$

21

Lesson 1: **Associative property**

- Understand and use the associative property of addition

Key words
- associative
- add
- sum
- total

Let's learn

What's the same about these two calculations?
What's different?
Which is more efficient? Why?

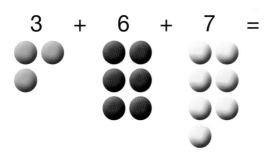　　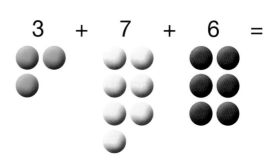

What about these two calculations?

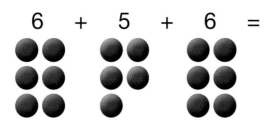　　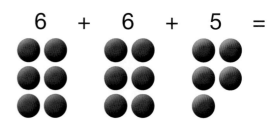

What is the best way to add these numbers? Explain why.

Guided practice

Write three examples to show the best way to add three numbers together by first making 10.

$4 + 6 + 7 = 17$	$8 + 2 + 5 = 15$	$5 + 5 + 9 = 19$

Lesson 2: **Addition and subtraction of multiples of 10**

Key words
* commutative
* inverse
* add
* sum
* total
* subtract
* difference

Number

* Add and subtract multiples of 10

Let's learn

Known number facts help us to add and subtract multiples of 10.

320 + 250 =   +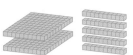

There are different ways to add these.

Method 1 32 + 25 = 57 +   =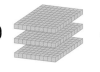

Then multiply 57 by 10 to give an answer of 570.

Method 2 3 + 2 = 5, so 300 + 200 = 500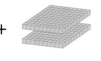

2 + 5 = 7, so 20 + 50 = 70
500 add 70 is 570.

We can do the same for subtraction.

Think about 250 – 120 =

Method 1 25 – 12 = 13

Then multiply 13 by 10 to give an answer of 130.

340 + 210 = 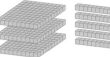 +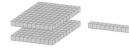

Use one of the methods above to work out the answer.
Then use another method to check.

Guided practice
Complete the calculations.

a 370 + 560 = [930]

b 560 – 420 = [140]

Lesson 3: Addition and subtraction with 3-digit numbers

Key words
- estimate
- add
- sum
- total
- subtract
- difference

- Estimate, add and subtract to and from 3-digit numbers

Let's learn

Partitioning and known number facts help us to add and subtract three-digit numbers.

Add and subtract 1s to and from a 3-digit number

$$465 + 8 = 400 + 60 + 5 + 8$$
$$= 400 + 60 + 13$$
$$= 400 + 73$$
$$= 473$$

$$523 - 6 = 500 + 23 - 6$$
$$= 500 + 17$$
$$= 517$$

Add and subtract 10s to and from a 3-digit number

$$376 + 50 = 300 + 70 + 6 + 50$$
$$= 300 + 120 + 6$$
$$= 420 + 6$$
$$= 426$$

$$634 - 70 = 630 + 4 - 70$$
$$= 560 + 4$$
$$= 564$$

Add and subtract 100s to and from a 3-digit number

$$283 + 500 = 200 + 80 + 3 + 500$$
$$= 700 + 80 + 3$$
$$= 783$$

$$715 - 400 = 700 + 10 + 5 - 400$$
$$= 300 + 10 + 5$$
$$= 315$$

What other strategies could you use to work out any of the above calculations? Which strategy is best? Explain why.

Guided practice

Estimate the answer to each calculation, writing your estimate in the bubble. Then work out the answer. Show all your working.

a 622 + 50 = (670)

$$6622 + 50 = 620 + 2 + 50$$
$$= 670 + 2$$
$$= 672$$

b 534 − 9 = (524)

$$534 - 9 = 534 - 10 + 1$$
$$= 524 + 1$$
$$= 525$$

Lesson 4: **More unknowns!**

- Use objects to represent unknown quantities

Key words
- add
- sum
- total
- subtract
- difference

Number

Let's learn

Remember! We can use any object to represent an unknown.

$20 -  = $16

The price of the pineapple is unknown.
We can use the relationship between addition and subtraction to work out the price.

$20 - $16 = $4   Let's check: $20 - (\$4) = $16
So the price of the pineapple is $4.

+ = $6

The price of one bunch of grapes is unknown.
We know that two bunches of grapes cost $6.
One bunch must cost half of $6.
One bunch of grapes costs $3.

Let's check:

$3 + $3 = $6

$100 -  = $25

What is the price of this pair of shoes?
Which facts help us to work this out?
If we know the cost of one pair of shoes – how much will two pairs cost?
What about three pairs?

Guided practice
Write the unknown number in each calculation.

a 50 + /30\ = 80   **b** 100 – (30) = 70

Number

Lesson 1: **Add 3-digit numbers and tens (A)**

• Estimate and add 3-digit numbers and tens

Let's learn

$$343 \quad + \quad 45 =$$

First, **estimate** the answer. $350 + 50 = 400$

Then look carefully at the numbers and decide whether to use a **mental strategy** or a **written method**.

Mental strategy

$$343 + 45 = 343 + 40 + 5$$
$$= 383 + 5$$
$$= 388$$

Written methods

Expanded methods		Formal method

```
   3 4 3        3 4 3          3 4 3
 +   4 5      +   4 5        +   4 5
   3 0 0            8          3 8 8
     8 0          8 0
         8      3 0 0
   3 8 8        3 8 8
```

Use a written method when a mental strategy cannot be used easily.

How are these three methods the same?
How are they different?

👥 Use Base 10 to make 315 and 61.
Estimate the sum of the two numbers.
Use a mental strategy to work it out.
Now practise using the three written methods.

Guided practice
Use an **expanded written method** and the **formal written method** to calculate $232 + 25$.

```
   2 3 2          2 3 2
 +   2 5        +   2 5
       7          2 5 7
     5 0
   2 0 0
   2 5 7
```

Lesson 2: **Add 3-digit numbers and tens (B)**

- Estimate and add 3-digit numbers and tens by regrouping

Key words
- add
- equals
- sum
- total
- estimate
- regroup
- expanded written method
- formal written method

Let's learn

$$447 \quad + \quad 45 =$$

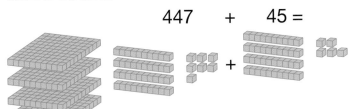

We could use a **mental strategy** to find the sum, but it might be more efficient to use a **written method**.

First, **estimate** the answer. 450 + 50 = 500

Now use a **written method**.

Expanded method

```
    4 4 7
+     4 5
      1 2
      8 0
    4 0 0
    4 9 2
```

Formal method

```
    4 4 7
+     4 5
    4 9 2
      1
```

Regroup the 12 ones into 1 ten and 2 ones.
Then add the 1 ten to the other tens.

How are these two calculations the same? How are they different?

Use Base 10 to make 315 and 69.

Estimate the sum of the two numbers.

Discuss whether you would use a mental strategy or a written method to find the sum. Explain why.

Each of you use your preferred method to work out the answer.

Which is the better strategy/method? Why?

Guided practice

Use an **expanded written method** and the **formal written method** to calculate 217 + 37.

```
    2 1 7            2 1 7
+     3 7        +     3 7
      1 4            2 5 4
      4 0              1
    2 0 0
    2 5 4
```

27

Lesson 3: **Add two 3-digit numbers (A)**

- Estimate and add 3-digit numbers by regrouping

Key words
- add
- equals
- sum
- total
- estimate
- regroup
- expanded written method
- formal written method

Let's learn

347 + 148 =

  +

First, **estimate** the answer. — 350 + 150 = 500

Then look carefully at the numbers and decide whether to use a **mental strategy** or a **written method**.

What mental strategy might you use? Why?

We could use a mental strategy, but for this calculation it might be more efficient to use a written method.

Expanded method

```
   3 4 7
+  1 4 8
   1 5
   8 0
 4 0 0
 4 9 5
```

Formal method

```
   3 4 7
+  1 4 8
   4 9 5
     1
```

Regroup the 15 ones into 1 ten and 5 ones.
Then add the 1 ten to the other tens.

Use Base 10 to make 235 and 148.

Estimate the sum of the two numbers.

Discuss whether you would use a mental strategy or a written method to find the sum. Explain why.

Each of you use your preferred method to work out the answer.

Which is the better strategy/method? Why?

Guided practice

Use an **expanded written method** and the **formal written method** to calculate 448 + 119.

```
   4 4 8
+  1 1 9
   1 7
   5 0
 5 0 0
 5 6 7
```

```
   4 4 8
+  1 1 9
   5 6 7
     1
```

Lesson 4: **Add two 3-digit numbers (B)**

- Estimate and add 3-digit numbers by regrouping

Key words
- add
- equals
- sum
- total
- estimate
- regroup
- expanded written method
- formal written method

Let's learn

$$467 + 172 =$$

First, **estimate** the answer. $470 + 170 = 640$

Then look carefully at the numbers and decide whether to use a **mental strategy** or a **written method**.

What mental strategy might you use? Why?

We could use a mental strategy, but for this calculation it might be more efficient to use a written method.

Expanded method	Formal method

```
    4 6 7
  + 1 7 2
        9
    1 3 0
    5 0 0
    6 3 9
```

```
    4 6 7
  + 1 7 2
    6 3 9
      1
```

Regroup the 13 tens into 1 hundred and 3 tens. Then add the 1 hundred to the other hundreds.

Use Base 10 to make 365 and 193.

Estimate the sum of the two numbers.

Discuss whether you would use a mental strategy or a written method to find the sum. Explain why.

Each of you use your preferred method to work out the answer.

Which is the better strategy/method? Why?

Guided practice

Use an **expanded written method** and the **formal written method** to calculate $493 + 182$.

```
    4 9 3
  + 1 8 2
        5
    1 7 0
    5 0 0
    6 7 5
```

```
    4 9 3
  + 1 8 2
    6 7 5
      1
```

Number

Lesson 1: **Subtract 3-digit numbers and tens (A)**

Key words
• **subtract**
• **equals**
• **difference**
• **estimate**

• Estimate and subtract 3-digit numbers and tens

Let's learn

286 – 54 =

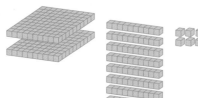

286 is the amount we have. 54 is the number we subtract.

First, **estimate** the answer.   300 – 50 = 250

Then look carefully at the numbers and decide whether to use a **mental strategy** or a **written method**.

Mental strategy

286 – 54 = 286 – 50 – 4
= 236 – 4
= 232

Written methods

Expanded method	Formal method

Use a written method when a mental strategy cannot be used easily.

Expanded method:
```
  200   80   6
–         50   4
  200   30   2
```

Formal method:
```
    2 8 6
 –    5 4
    2 3 2
```

How are these methods the same?
How are they different?

Use Base 10 to make 378.
Estimate the difference between 378 and 46.
Use a mental strategy to subtract 46 from 378.
Now practise using the two written methods.

Guided practice

Use the **expanded written method** and the **formal written method** to calculate 376 – 33

```
  300   70   6
–         30   3
  300   40   3
```

```
    3 7 6
 –    3 3
    3 4 3
```

Number

Lesson 2: **Subtract 3-digit numbers and tens (B)**

- Estimate and subtract 3-digit numbers and tens by regrouping

Let's learn

486 – 57 =

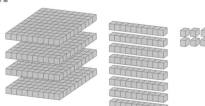

We could use a **mental strategy** to find the **difference**, but it might be more efficient to use a **written method**.

First, **estimate** the answer.   500 – 50 = 450

Now use a **written method**.

1 ten is regrouped into 10 ones. There are now 7 tens and 16 ones.

Expanded method			
	70	16	
400	8̶0̶	6̶	
–		50	7
400		20	9

Formal method	
	7 1
4 8̶ 6
– 5 7
4 2 9

How are these methods the same?
How are they different?

Use Base 10 to make 384.

Estimate the difference between 384 and 66.

Discuss whether you would use a mental strategy or a written method to find the difference. Explain why.

Each of you use your preferred method to work out the answer.

Which is the better strategy/method? Why?

Guided practice

Use the **expanded written method** and the **formal written method** to calculate 194 – 56.

		80	14	
100	9̶0̶	4̶	
–		50	6
100	30	8	

	8 1	
1 9̶ 4
– 5 6
1 3 8

Lesson 3: **Subtract two 3-digit numbers (A)**

- Estimate and subtract 3-digit numbers by regrouping

Key words
- subtract
- equals
- difference
- estimate
- regroup

Let's learn

385 – 138 =

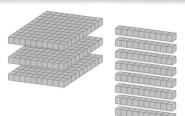

First, **estimate** the answer.

400 – 150 = 250

Then look carefully at the numbers and decide whether to use a **mental strategy** or a **written method**.

What mental strategy might you use? Why?

We could use a **mental strategy**, but for this calculation it might be more efficient to use a **written method**.

Expanded method	Formal method

$$\begin{array}{rrr} & ^{70} & ^{15} \\ 300 & \cancel{80} & \cancel{5} \\ -\ 100 & 30 & 8 \\ \hline 200 & 40 & 7 \end{array}$$

$$\begin{array}{r} ^{7}\ ^{1} \\ 3\,\cancel{8}\,5 \\ -\ 1\ 3\ 8 \\ \hline 2\ 4\ 7 \end{array}$$

1 ten is regrouped into 10 ones.
There are now 7 tens and 15 ones.

Use Base 10 to make 235.

Estimate the difference between 235 and 117.

Discuss whether you would use a mental strategy or a written method to find the difference. Explain why.

Each of you use your preferred method to work out the answer.

Which is the better strategy/method? Why?

Guided practice

Use the **expanded written method** and the **formal written method** to calculate 263 – 147.

$$\begin{array}{rrr} & ^{50} & ^{13} \\ 200 & \cancel{60} & \cancel{3} \\ -\ 100 & 40 & 7 \\ \hline 100 & 10 & 6 \end{array}$$

$$\begin{array}{r} ^{5}\ ^{1} \\ 2\,\cancel{6}\,3 \\ -\ 1\ 4\ 7 \\ \hline 1\ 1\ 6 \end{array}$$

Number

Lesson 4: **Subtract two 3-digit numbers (B)**

- Estimate and subtract 3-digit numbers by regrouping

Key words
- minuend
- subtract
- subtrahend
- equals
- difference
- estimate
- regroup

Let's learn

$437 - 284 =$

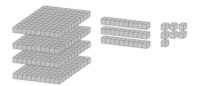

First, **estimate** the answer. $440 - 280 = 160$

Then look carefully at the numbers and decide whether to use a **mental strategy** or a **written method**.

What mental strategy might you use? Why?

We could use a **mental strategy**, but for this calculation it might be more efficient to use a **written method**.

Expanded method	Formal method	
300 130	3 1	1 hundred is regrouped into 10 tens. There are now 3 hundreds and 13 tens.

Expanded method

$$\begin{array}{r} \overset{300}{\cancel{400}}\quad \overset{130}{\cancel{30}}\quad 7 \\ -\ 200\quad\ \ 80\quad 4 \\ \hline 100\quad\ \ 50\quad 3 \end{array}$$

Formal method

$$\begin{array}{r} \overset{3}{\cancel{4}}\overset{1}{3}\,7 \\ -\ 2\,8\,4 \\ \hline 1\,5\,3 \end{array}$$

1 hundred is regrouped into 10 tens.
There are now 3 hundreds and 13 tens.

Use Base 10 to make 526.

Estimate the difference between 526 and 163.

Discuss whether you would use a mental strategy or a written method to find the difference. Explain why.

Each of you use your preferred method to work out the answer.

Which is the better strategy/method? Why?

Guided practice

Use the **expanded written method** and the **formal written method** to calculate $453 - 172$.

$$\begin{array}{r} \overset{300}{\cancel{400}}\quad \overset{150}{\cancel{50}}\quad 3 \\ -\ 100\quad\ \ 70\quad 2 \\ \hline 200\quad\ \ 80\quad 1 \end{array}$$

$$\begin{array}{r} \overset{3}{\cancel{4}}\overset{1}{5}\,3 \\ -\ 1\,7\,2 \\ \hline 2\,8\,1 \end{array}$$

Lesson 1: **Multiplication and division**

- Understand the relationship between multiplication and division

Let's learn

There are 15 counters in this array…

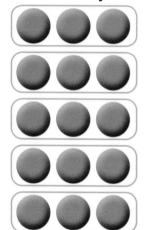

…and there are 15 counters in this array.

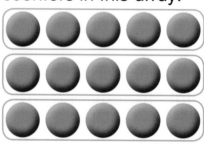

In this array, there are 3 groups of 5 counters.

In this array, there are 5 groups of 3 counters.

The arrays show these calculations:

$3 \times 5 = 15$ $5 \times 3 = 15$

$15 \div 3 = 5$ $15 \div 5 = 3$

 Use counters to make an array with 5 rows of 4 counters.
Write the two multiplication calculations.
Write the two division calculations.

Guided practice

Write a multiplication and a division for each array.

 $2 \times 1 = 2$
$2 \div 1 = 2$

$2 \times 2 = 4$
$4 \div 2 = 2$

 $2 \times 4 = 8$
$8 \div 2 = 4$

Lesson 2: **Checking multiplication and division**

Number

Key words
- **multiplication**
- **product**
- **division**
- **quotient**
- **inverse**
- **array**
- **groups**

- Use the relationship between multiplication and division

Let's learn

We can use division to check multiplication.
We can use multiplication to check division.

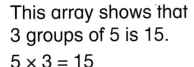

This array shows that
3 groups of 5 is 15.
5 × 3 = 15

This shows that we can make
3 groups of 5 from 15.
15 ÷ 5 = 3

Use counters to create an array to show that 5 × 4 = 20.
Then take away groups of 5 counters.
How many groups of 5 did you make?
Write the division calculation.

Guided practice

Write the division calculation you can use to check each multiplication.

2 × 3 = 6   6 ÷ 2 = 3   2 × 8 = 16   16 ÷ 2 = 8

Write the multiplication calculation you can use to check each division.

70 ÷ 10 = 7   10 × 7 = 70   45 ÷ 5 = 9   5 × 9 = 45

Lesson 3: **Commutativity**

- Understand that multiplication is commutative

Let's learn

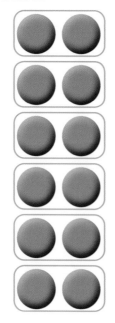

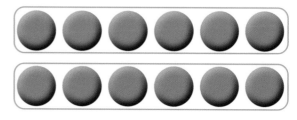

This array shows $6 \times 2 = 12$.

This array shows $2 \times 6 = 12$.

When we multiply two numbers, it doesn't matter which way around we do it. Multiplication is commutative.

The answer will always be the same.

Use large squared paper.
Draw dots to make any array you wish.
Cut out the array.

Write the two multiplication facts for your array.

Guided practice

Write the multiplications for each array.

| $3 \times 2 = 6$ | $2 \times 3 = 6$ | $5 \times 2 = 10$ | $2 \times 5 = 10$ |

Number

Number

Lesson 4: **Using place value to multiply**

• Understand that multiplication is distributive

Let's learn

15 × 2 =

Partitioning numbers into tens and ones can make multiplication easier.

Partition 15 into 10 and 5.

15 × 2

10 × 2 = 20

Then multiply each number by 2.

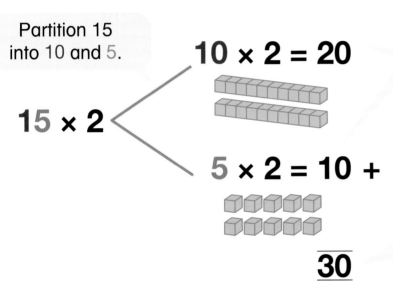

5 × 2 = 10 +

30

Then add the two products to get the answer.

Use Base 10 equipment to show the calculation 17 × 2.

Group the tens together. Group the ones.

Now find the answer.

Write what you have done in numbers.

Guided practice

a 18 × 3 =

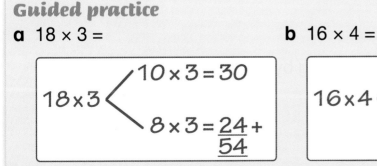

18×3
$10 \times 3 = 30$
$8 \times 3 = \underline{24} + \underline{54}$

b 16 × 4 =

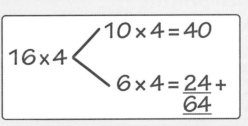

16×4
$10 \times 4 = 40$
$6 \times 4 = \underline{24} + \underline{64}$

Lesson 1: **5 and 10 times tables**

Number

- Understand the relationship between the 5 and 10 times tables

Key words
- multiply
- product
- divide
- times table
- equivalent

Let's learn

These bars show that 10 has the same value as two 5s.

It shows that if we halve 10 we get 5.

Any number can be multiplied by 5 by first multiplying it by 10 and then halving.

5	5

10

$$14 \times 5 =$$

$$14 \times 10 = 140$$

First **multiply** 14 **by 10**

$$140 \div 2 = 70$$

Then **halve the product** to get the answer.

Choose a number between 10 and 20. Make sure your number is different from your partner's.

Use the method you have learned to multiply your number by 5.

Explain to your partner what you have done.

What's the same about your multiplication and your partner's?

What's different?

Guided practice

Work out the product by multiplying by 10 and halving.

$$12 \times 5 =$$

$$12 \times 10 = 120$$
$$120 \div 2 = 60$$

Lesson 2: **2, 4 and 8 times tables**

Key words
* **multiply**
* **product**
* **divide**
* **times table**
* **equivalent**

* Understand the relationship between the 2, 4 and 8 times tables

Let's learn

You already know the 2 times table.

You can use the 2 times table and doubling to help you work out the answers to the 4 times table and the 8 times table.

Remember! Doubling is the same as multiplying by 2.

Use 2 × 7 = 14, to work out **4 × 7 =**

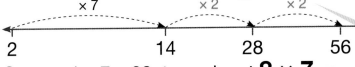

× 7　　　　　× 2

2　　　　14　　　　28

For the 4 times table, double the answer to the 2 times table.

Use 2 × 7 = 14, to work out **8 × 7 =**

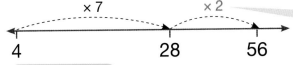

× 7　　　　× 2　　　× 2

2　　　14　　　28　　　56

For the 8 times table, double the answer to the 2 times table, …

Or, use 4 × 7 = 28, to work out **8 × 7 =**

× 7　　　　× 2

4　　　28　　　56

… then double that answer.

For the 8 times table, double the answer to the 4 times table.

Remember! If you know the multiplication fact, you also know the related division fact.

4 × 7 = 28　So, 28 ÷ 4 = 7　　　　8 × 7 = 56　So, 56 ÷ 8 = 7

Choose a number less than 10. Make sure your number is different from your partner's.

Use doubling to multiply it by 4 and by 8.

Explain to your partner what you have done.

What's the same about your multiplications and your partner's? What's different?

Guided practice
Complete each times table fact.

a 4 × 7 = | 28 |　**b** 8 × 5 = | 40 |　**c** 12 ÷ 4 = | 3 |　**d** 80 ÷ 8 = | 10 |

Lesson 3: **Multiples**

- Recognise multiples of 2, 5 and 10 (up to 1000)

Key words
- **multiple**
- **product**

Let's learn

Multiples are the products of the multiplication tables.

This grid shows the first ten multiples of 2, 5 and 10.

5, 10, 15, 20, 25, 30, 35, 40, 45 and 50 are **multiples of 5**.

×	2	5	10
1	2	5	10
2	4	10	20
3	6	15	30
4	8	20	40
5	10	25	50
6	12	30	60
7	14	35	70
8	16	40	80
9	18	45	90
10	20	50	100

10, 20, 30, 40, 50, 60, 70, 80, 90 and 100 are **multiples of 10**.

2, 4, 6, 8, 10, 12, 14, 16, 18 and 20 are **multiples of 2**.

What other numbers are multiples of 2?
What about multiples of 5?
What about multiples of 10?

Guided practice

Which is the odd one out in this set? Why?

15, 30, 25, 16, 10　　16

All the other numbers are multiples of 5.

Number

Lesson 4: **Counting in steps**

• Count in steps of the same size

Let's learn

5 × 1 ●●●●●

5 × 2 ●●●●● ●●●●●

5 × 3 ●●●●● ●●●●● ●●●●●

5 × 4 ●●●●● ●●●●● ●●●●● ●●●●●

5 × 5 ●●●●● ●●●●● ●●●●● ●●●●● ●●●●●

This shows a pattern of counting in 5s.

It also shows the multiplication facts for the 5 times table.

We can link counting in steps of a number to multiplication.

Counting in 2s links to the 2 times table.

Counting in 4s links to the 4 times table.

Counting in 8s links to the 8 times table.

With your partner, continue the pattern to 5 × 10.

Write down the times tables facts you have made.

Are those the only times tables facts for 5?

If you think they are, tell your partner why.

What do you notice about the ones digit in this times table?

Guided practice

Which two numbers come next in this sequence?

24, 28, 32, 36, 40, 44 , 48

Why?

Because this sequence increases by 4 each time.

41

Number

Lesson 1: **3 and 6 times tables**

- Understand the relationship between the 3 and 6 times tables

Key words
- times table
- multiply
- multiple
- product

Let's learn

You already know the 3 times table.

You can use the 3 times table and doubling to help you work out the answers to the 6 times table.

Remember! Doubling is the same as multiplying by 2.

Use $3 \times 7 = 21$, to work out **$6 \times 7 =$**

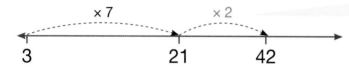

For the 6 times table, double the answer to the 3 times table.

You can also use the 2 times table and tripling to help you work out the answers to the 6 times table. Tripling is the same as multiplying by 3.

Use $2 \times 4 = 8$, to work out **$6 \times 4 =$**

For the 6 times table, triple the answer to the 2 times table.

Remember! If you know the multiplication fact, you also know the related division fact.

$6 \times 7 = 42$ So, $42 \div 6 = 7$ $6 \times 4 = 24$ So, $24 \div 6 = 4$

We can make a rule that says we can double 3 times tables facts to give us facts for the 6 times table.

What are the multiples of 6 that can be made from these multiples of 3?

3   12   18   21   27   30

Guided practice
Complete each times table fact.

a $6 \times 7 = \boxed{42}$ **b** $6 \times 5 = \boxed{30}$ **c** $12 \div 6 = \boxed{2}$ **d** $6 \div 6 = \boxed{1}$

Lesson 2: **9 times table**

* Know the 9 times table

Key words
* **multiply**
* **product**
* **divide**
* **quotient**
* **times table**

Number

Let's learn

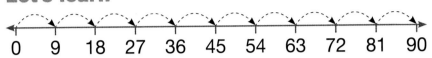

$$0 \quad 9 \quad 18 \quad 27 \quad 36 \quad 45 \quad 54 \quad 63 \quad 72 \quad 81 \quad 90$$

We can count in steps of 9 and link the numbers to the 9 times table.
The second number in the count is 18. This is the same as $9 \times 2 = 18$.
The sixth number in the count is 54. This is the same as $9 \times 6 = 54$.

We know the commutative facts for the 1, 2, 3, 4, 5, 6, 8 and 10 times tables, so we already know:

$9 \times 1 = 9$ $9 \times 2 = 18$ $9 \times 3 = 27$ $9 \times 4 = 36$

$9 \times 5 = 45$ $9 \times 6 = 54$ $9 \times 8 = 72$ $9 \times 10 = 90$

You can also use the 3 times table and tripling to help you work out the answers to the 9 times table.

Remember! Tripling is the same as multiplying by 3.

Use $3 \times 4 = 12$, to work out **9 × 4 =**

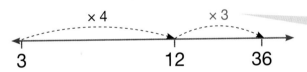

$$3 \qquad\qquad 12 \qquad\qquad 36$$

For the 9 times table, triple the answer to the 3 times table.

Remember! If you know the multiplication fact, you also know the related division fact.

$9 \times 4 = 36$ So, $36 \div 9 = 4$

Explain to your partner why we can triple the 3 times table facts to make the 9 times table facts.

Guided practice

Show how you can use the 3 times table to work out the answer to the 9 times table fact.

$9 \times 7 = \boxed{63}$ $\begin{array}{l} 3 \times 7 = 21 \\ \qquad\qquad\quad \times 3 \\ 9 \times 7 = 63 \end{array}$

43

Lesson 3: **Multiplication and division facts (1)**

- Know the 1, 2, 3, 4, 5, 6, 8, 9, and 10 times tables

Key words
- multiply
- product
- divide
- quotient
- times table
- equivalent

Let's learn

×	1	2	3	4	5	6	7	8	9	10
1	1	2	3	4	5	6	7	8	9	10
2	2	4	6	8	10	12	14	16	18	20
3	3	6	9	12	15	18	21	24	27	30
4	4	8	12	16	20	24	28	32	36	40
5	5	10	15	20	25	30	35	40	45	50
6	6	12	18	24	30	36	42	48	54	60
7	7	14	21	28	35	42	49	56	63	70
8	8	16	24	32	40	48	56	64	72	80
9	9	18	27	36	45	54	63	72	81	90
10	10	20	30	40	50	60	70	80	90	100

What do you notice about the numbers in

- the red columns?
- the green columns?
- the blue columns?

21
3   7

36
9   4

24
8   3

Draw some trios.
Write the two multiplication and division facts.

Guided practice

Write two multiplication facts that have a product of 18. Write the corresponding division facts.

$3 \times 6 = 18 \quad 9 \times 2 = 18$ $18 \div 3 = 6 \quad 18 \div 9 = 2$

Number

Lesson 4: **Multiplication and division facts (2)**

- Know the 1, 2, 3, 4, 5, 6, 8, 9, and 10 times tables

Let's learn

The times tables can be represented in different ways.

Using a number line.

Step count along the number line to see that $6 \times 9 = 54$

Remember!

- Multiplication is commutative so if we know $6 \times 9 = 54$, we also know that $9 \times 6 = 54$
- Division is the inverse of multiplication so we also know: $54 \div 6 = 9$ and $54 \div 9 = 6$

Commutative means 'can be done in any order'.

Using a multiplication grid.

×	4
1	4
2	8
3	12
4	16
5	20
6	24
7	28
8	32
9	36
10	40

$5 \times 4 = 20$
$4 \times 5 = 20$
$20 \div 5 = 4$
$20 \div 4 = 5$

Using a trio.

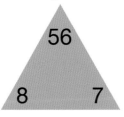

Choose a times table and draw a number line similar to the 6 times table number line in Let's learn.

Guided practice

Starting at zero, count on in steps of 3 seven times. What number do you say?

21

Write the two multiplication and two division facts.

$3 \times 7 = 21 \quad 7 \times 3 = 21$

$21 \div 3 = 7 \quad 21 \div 7 = 3$

Lesson 1: **Multiplying by repeated addition**

Key words
- **multiplication**
- **multiplicand**
- **multiplied by**
- **multiplier**
- **product**

- Estimate and multiply 2-digit numbers by 2, 3, 4 and 5

Number

Let's learn

Multiplication is the same as repeated addition.

We can use repeated addition to solve $35 \times 3 =$
$35 + 35 + 35 = 105$

A number line can help with this.

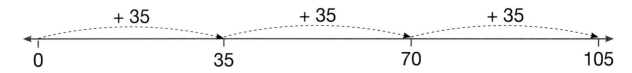

We can use mental calculation strategies.
- To multiply by 2, use doubling.
- To multiply by 4, double and then double again.
- To multiply by 5, multiply by 10 and halve the result.

We must remember to estimate the product first and check that our answer is similar to our estimate.

Talk to your partner about the simplest way to work out 48×5.
Write down some 2-digit numbers less than 50.
Multiply the numbers by 5 by multiplying by 10 and halving.
Now try halving your numbers and multiplying by 10.
Are your answers the same?

Guided practice

What mental calculation strategy would you use for multiplying by 5?
Give an example.

Multiply by 10 and halve the product.

$14 \times 10 = 140$ and half of $140 = 70$

Lesson 2: **Multiplying with arrays**

- Estimate and multiply 2-digit numbers by 2, 3, 4 and 5

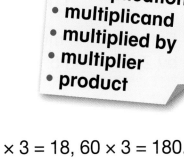

Key words
- multiplication
- multiplicand
- multiplied by
- multiplier
- product

Let's learn

We can use repeated addition and mental calculations to multiply numbers.

64 × 3 = We must estimate the product first. $6 \times 3 = 18$, $60 \times 3 = 180$. 180 is a good estimate.

We can use a number line.

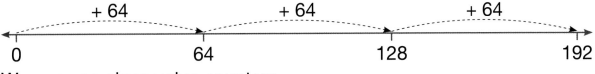

We can use place value counters.

There are three groups of 4 ones. That makes 12 altogether.

Exchange 10 ones for 1 ten. That makes 19 tens.

We exchange 10 tens for 1 hundred.
This is our product:

$$100 + 90 + 2 = 192$$

Make an array to show 27 × 3 =.

Talk to your partner about how you would work out the product.

Guided practice

How would you use this array to multiply 34 by 3?

$3 \text{ tens} \times 3 = 90$
$4 \text{ ones} \times 3 = 12$
$90 + 12 = 102$

Lesson 3: **Multiplying by the grid method**

Number

- Estimate and multiply 2-digit numbers by 2, 3, 4 and 5

Let's learn

Using place value counters can be helpful when multiplying.

These place value counters show the calculation 43×5.

This grid also shows $43 \times 5 =$

×	40	3
5	200	15

$$\begin{array}{r} 2\ 0\ 0 \\ +\ 1\ 5 \\ \hline 2\ 1\ 5 \end{array}$$

We must estimate the product first. $4 \times 5 = 20$, $40 \times 5 = 200$. 200 is a good estimate.

Talk to your partner about the two methods in Let's learn. What's the same? What's different?

How does the grid show how to calculate the product?

Guided practice

Draw a grid and work out $21 \times 5 =$

×	20	1
5	100	5

$$\begin{array}{r} 1\ 0\ 0 \\ +\ \ \ 5 \\ \hline 1\ 0\ 5 \end{array}$$

Lesson 4: **Multiplying by partitioning**

- Estimate and multiply 2-digit numbers by 2, 3, 4 and 5

Key words
- multiplication
- multiplicand
- multiplied by
- multiplier
- product

Let's learn

Partitioning is another way to multiply 2-digit numbers by 2, 3, 4 and 5.

Look at this calculation: 34 × 5

First, estimate the product.　Between 150 and 200.

Partition the multiplicand into tens and ones.

Multiply the tens by the multiplier.

$$34 \times 5$$

Then multiply the ones by the multiplier.

$$30 \times 5 \qquad 4 \times 5$$

Finally, add the two numbers together to give the final product.

$$150 \quad + \quad 20 \quad = 170$$

To check, multiply the product by 10 and halve it.
34 × 10 = 340. Half of 340 = 170.

Think of a 2-digit number. Multiply it by 5.
First, estimate the product.
Next, use the mental calculation strategy of multiplying by 10 and halving.
Finally, use partitioning to check your answer.
Were both your products the same?
Were your products close to your estimate?

Guided practice
Use partitioning to multiply 35 by 3.

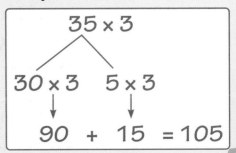

$$35 \times 3$$
$$30 \times 3 \qquad 5 \times 3$$
$$90 \quad + \quad 15 \quad = 105$$

Lesson 1: **Dividing using known facts**

* Estimate and divide 2-digit numbers by 2, 3, 4 and 5

Key words
* multiplication
* division
* dividend
* divisor
* quotient

Let's learn

There are lots of ways to work out division calculations.

$$35 \div 5 =$$

Division is the same as repeated subtraction.

To divide 35 by 5, keep subtracting 5.

$$35 - 5 - 5 - 5 - 5 - 5 - 5 - 5$$

5 has been subtracted 7 times.

Use multiplication facts and count on in 5s up to 35.

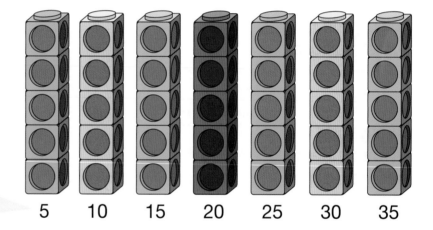

There are 7 groups of 5.

| 5 | 10 | 15 | 20 | 25 | 30 | 35 |

Use mental calculation strategies.

To divide by **2**: Use halving.

To divide by **4**: Halve and halve again.

To divide by **5**: Double then divide by 10.

👥 Talk to your partner about how to work out $70 \div 5 =$. Estimate the quotient first. Now think of some 2-digit multiples of 10.

Divide them by 5 by doubling and dividing by 10.

Guided practice

Solve this division by halving and halving again.

$$52 \div 4 = \boxed{13}$$

$$52 \div 2 = 26$$
$$26 \div 2 = 13$$

Lesson 2: **Dividing by partitioning**

Key words
* multiplication
* division
* dividend
* divisor
* quotient

- Estimate and divide 2-digit numbers by 2, 3, 4 and 5

Let's learn

We can use the times tables facts for 2, 3, 4 and 5 to help us divide.

$$75 \div 5 =$$

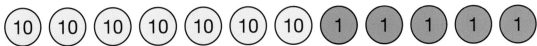

We need to estimate the quotient first. We know that it is more than 10. An estimate could be between 10 and 20. We can partition numbers into tens and another number.

Partition 75 into 50 and 25

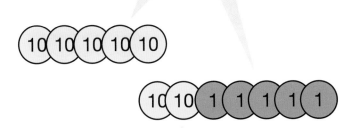

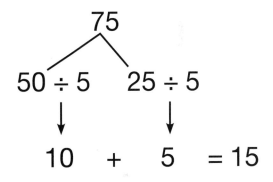

Divide 50 and 25 by 5. Then add 10 and 5 to find the quotient.

Talk to your partner about how to partition 75.

Can you partition 75 into 40 and 35 and divide by 5?

What about 30 and 45?

What about 20 and 55?

What about 10 and 65?

Which is the easiest way to divide by 5? Why?

Guided practice

Draw place value counters to show how to partition 32 in three ways.

(10) (10) (10)   (1) (1)
    30   2

(10) (10)   (10) (1) (1)
    20   12

(10)   (10) (10) (1) (1)
  10   22

Lesson 3: **Dividing with arrays**

- Estimate and divide 2-digit numbers by 2, 3, 4 and 5

Key words
- division
- dividend
- divisor
- quotient
- exchange

Let's learn

65 ÷ 5 =

We can use place value counters in a grid to help us divide. An estimate for the quotient could be between 10 and 20.

Step 1

Partition 65 into tens and ones.

10s | 1s

Step 2

We can make 1 group of 5 tens.

10s | 1s

Step 3

We exchange the extra ten for 10 ones.

10s | 1s

Step 4

Now we make 3 groups of 5 ones.

10s | 1s

Step 5 65 ÷ 5 is: 1 group of 5 tens
and 3 groups of 5 ones.

so, 65 ÷ 5 = 13

Look at this calculation: 75 ÷ 5 =
How is this the same as 65 ÷ 5 =? How is it different?
How would you work out the quotient?

Guided practice
Look at this grid.
Draw the groups we make if we divide this number by 3.
What is the answer? | 13 |

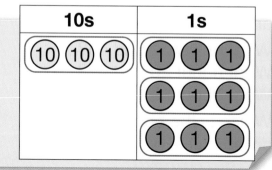

10s | 1s

Lesson 4: **Division with remainders**

- Estimate and divide 2-digit numbers by 2, 3, 4 and 5

Number

Let's learn

We can write divisions by using the division bracket. It looks like this.

$$92 \div 4 =$$

division bracket

2 3 ← quotient

divisor → 4) 9 ^{1}2 ← dividend

We always use a division bracket when we use a written method to divide. We always estimate the quotient first. An estimate might be about 20 because we know that $20 \times 4 = 80$.

Some divisions can have a remainder like this calculation:

$$68 \div 5 =$$

1 3   remainder 3

5) 6 ^{1}8

Remember!

We should always check our calculations.

To check a division, multiply the quotient by the divisor.

Add the two products together.

If there is a remainder, add that on.

$10 \times 5 = 50$

$3 \times 5 = 15$

$50 + 15 = 65$

$65 + 3 = 68$

Explain to your partner how to divide 68 by 5, using place value counters.

Explain how you would record using the written method.

Guided practice

Use the written method to answer the division.

2 3 remainder 2

3) 7 ^{1}1

Lesson 1: **Writing money**

• Write money with the decimal point

Number

Key words
• coins
• notes
• decimal point
• place holder

Let's learn

When we buy or sell things, we use coins and notes.

If we use dollars and cents, we write a dot to separate the dollars and cents.

We write the number of **dollars before** the dot and the number of **cents after** the dot.

The dot is called a decimal point.

This amount totals three dollars and 47 cents.
We write this as $3.47

This amount totals one dollar
and three cents.
We write this as $1.03

There are no tens of cents, so we use 0 as a place holder in the first position after the dot.

How much money is shown here?
Talk to your partner about the best way to find out.
How will you write this amount?

Guided practice

How much money is this?
Write the amount in dollars $ and cents.

$ 11.16

Number

Lesson 2: **Finding totals**

- Add amounts of money to find totals

Key words
- **total**
- **spend**

Let's learn

$4.00

$12.00

Rasheed buys a T-shirt and a cap.

How much does he spend?

$12 + $4 = $16

$16.50

$23

Aarav buys a pair of shorts and shoes.
How much does he spend?

Add the **dollars** together.

$23 + $16 = $39

Then **add** the cents.
$39.50

What is a good strategy to use to add $4.50 and $3.00 together?
Talk to your partner about which strategy you would use and why.

Guided practice
Alison buys two tops costing $14 and $17.
How much does Alison spend?
She spends $31.

Lesson 3: **Finding change**

- Subtract amounts of money to find change

Key words
- **change**
- **subtract**
- **difference**

Let's learn

Rainbow restaurant

Pasta	Regular	Large		Sides	Small	Medium
Macaroni cheese	$4.50	$7.00		Three bean salad	$1.40	$2.50
Spaghetti bolognese	$4.50	$7.00		Tomato salsa	$1.40	$2.50
Carbonara	$4.50	$7.00		Leafy greens	$1.40	$2.50

Grilled	Regular	Large		Desserts	2 scoops	4 scoops
Grilled chicken	$6.00	$10.00		Ice cream	$1.00	$1.50
Tuna steak	$8.00	$12.00		Fruit salad	$2.25	
Vegetable kebab	$5.00	$8.00		Banana sundae	$3.00	

If you buy a regular Grilled chicken, what will be the change from $10?

$10 − £6 = $4

A good strategy to work out the change is to find the **difference** by counting on from the cost of the item to the amount of money offered.

If you buy a large Carbonara and a small Tomato salsa, what will be the change from $20?

$7 + $1.40 = $8.40

$20 − £8.40 = $11.60

First find the **total**, then work out the **difference**.

With your partner, choose a main dish, a side and a dessert.
Work out the total cost.
Now work out your change from $50.

Guided practice

Razza has $20. He buys a football.
How much change does he get?

$8

He gets $12 change.

Lesson 4: **Solving problems with money**

• Solve problems with money

Number

Key words
• total
• change

Let's learn

Remember! We use addition to find totals and subtraction to find change.

The total cost of an elephant and a snake is $28.

The change from $50 is $22.

Sometimes we need to multiply and divide to find totals.

To find the cost of three toy gorillas, we multiply $20 by 3 which is $60.

If two lions cost $18, we know that one must cost $9.

Look at the Cuddly Toy Price List. Pick three toys to buy.

Work with your partner to find the total cost.

Then work out the change from $100.

Cuddly Toy Price List	
lion	$9
giraffe	$10
elephant	$15
snake	$13
gorilla	$20
polar bear	$11
seal	$9
monkey	$18

Guided practice

Cinema tickets cost $6. Children's tickets are half price.

How much does it cost for 2 adults and 1 child to go to the cinema?

Show your working.

I double $ 6 for the two adult tickets. This is $ 12.

I halve $ 6 for one child ticket. This is $ 3.

Then I add $ 12 and $ 3. The total cost is $ 15.

Lesson 1: **Understanding place value (A)**

- Understand the value of each digit in a 3-digit number
- Compose and decompose 3-digit numbers, using hundreds, tens and ones

Key words
- **hundreds**
- **tens**
- **ones**
- **position**
- **add**

Let's learn

This is a place value chart.

100s	10s	1s
3	6	4

The digit 3 is in the hundreds position.
The digit 6 is in the tens position.
The digit 4 is in the ones position.

To find the value of each digit we look at its position in the place value chart.

The 3 represents 3 hundreds = 300.
The 6 represents 6 tens = 60.
The 4 represents 4 ones = 4.

To find the whole number, we add the values together.
300 + 60 + 4 = 364

We can also show 364 using Base 10 equipment.

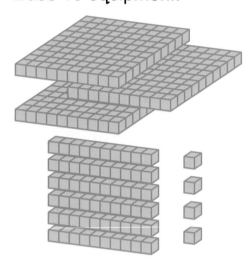

Talk to your partner about the number 534.
What positions are the digits in?
What are their values?
What will you do to make the whole number?

Guided practice
Write the value of each digit in these numbers.

248 = ☐2☐ hundreds + ☐4☐ tens + ☐8☐ ones

629 = ☐6☐ hundreds + ☐2☐ tens + ☐9☐ ones

Lesson 2: **Understanding place value (B)**

> **Key words**
> • **hundreds**
> • **tens**
> • **ones**
> • **position**
> • **place**
> • **add**
> • **zero**
> • **place holder**

- Understand the value of each digit in a 3-digit number

Let's learn

We can show 3-digit numbers in different ways.

Place value chart

100s	10s	1s
4	0	3

Abacus

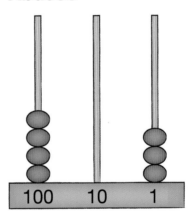

100   10   1

Place value counters

$$403 = 400 + 0 + 3$$

Make up some numbers that have a zero in the tens position.
Now make up some numbers that have a zero in the ones position.
Tell your partner what the zero does.

Guided practice

Write the value of each digit in these numbers.

a $914 = \boxed{900} + \boxed{10} + \boxed{4}$

b $468 = \boxed{400} + \boxed{60} + \boxed{8}$

Lesson 3: **Regrouping**

- Understand that numbers can be regrouped in different ways

Key words
- **hundreds**
- **tens**
- **ones**
- **regroup**

Let's learn

This is 143 in groups of 100, 10 and 1.

            100 + 40 + 3 = 143

We can regroup 143 as 143 ones.
We can also regroup 143 as 14 tens and 3 ones.

There are lots of ways to regroup 143, for example:

- 1 hundred, 3 tens and 13 ones
- 13 tens and 13 ones
- 12 tens and 23 ones
- 11 tens and 33 ones

Think of as many ways as you can to regroup this number.
Start with ones.
Then look at tens and ones.

Guided practice

How is 234 represented?

a

2 hundreds , 3 tens and 4 ones

b

1 hundred , 13 tens and 4 ones

c

23 tens and 4 ones

Lesson 4: **Comparing numbers**

- Compare 3-digit numbers

Key words
- compare
- greater than
- less than

Number

Let's learn

When comparing 3-digit numbers, start by looking at the hundreds digits.

456

456 has 4 hundreds

456 **is greater than** 378

456 **>** 378

378

378 has 3 hundreds

378 **is less than** 456

378 **<** 456

If the hundreds digits are the same, then look at the tens digits.

456

456 has 4 hundreds
456 has 5 tens

456 **is less than** 478

456 **<** 478

478

478 has 4 hundreds
478 has 7 tens

478 **is greater than** 456

478 **>** 456

Each choose a 3-digit number. Write your numbers side by side on a piece of paper.

Now write the correct symbol between them.

Do this two more times.

Guided practice

Write the correct symbol (> or <) between each pair of numbers.

a 546 $\boxed{>}$ 480   **b** 247 $\boxed{<}$ 271

Lesson 1: **Ordering numbers**

• Order 3-digit numbers

Key words
- **hundreds**
- **tens**
- **ones**
- **position**
- **order**
- **ascending**
- **descending**

Let's learn

When we order numbers, we order from smallest to largest – ascending order, or largest to smallest – descending order.

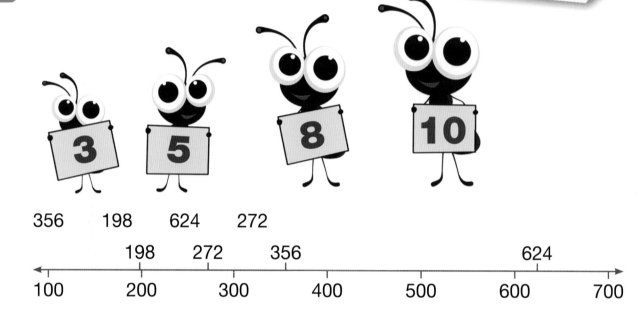

The order from smallest to largest is 198, 272, 356, 624.

The order from largest to smallest is 624, 356, 272, 198.

👥 Talk to your partner about how you would order 465, 132, 463, 182.
Write the numbers in order, from smallest to largest.
Then write them in order, from largest to smallest.

Guided practice
Write these numbers in order, from smallest to largest.
356, 456, 156, 756
156, 356, 456, 756

Lesson 2: **Multiplying by 10**

- Use place value to multiply 2-digit numbers by 10

Number

Let's learn

When we multiply by 10, the value of each digit becomes 10 times larger.

100s	10s	1s
	6	3
6	3	0

The 6 tens become 6 hundreds.

The 3 ones become 3 tens.

We need to write 0 as a placeholder in the ones position.

$$63 \times 10 = 630$$

Write five different 2-digit numbers.

Multiply each number by 10. Explain to your partner what happens when you multiply them by 10.

Guided practice

Complete these calculations.

a $38 \times 10 = \boxed{380}$

b $57 \times 10 = \boxed{570}$

c $61 \times 10 = \boxed{610}$

d $70 \times 10 = \boxed{700}$

Number

Lesson 3: **Rounding to the nearest 10**

Key words
- **hundreds**
- **tens**
- **ones**
- **position**
- **estimate**
- **round**
- **round up**
- **round down**

- Round 3-digit numbers to the nearest 10

Let's learn

If we need to work out 369 + 223, the first thing to do is to estimate the answer.

Rounding is a useful strategy to use for **estimating**.

Remember!

round **down** 0 1 2 3 4   5 6 7 8 9 round **up**

369 rounds **up** to 370.

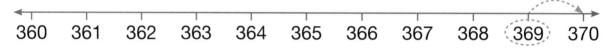

223 rounds **down** to 220.

An estimate of 369 + 223 is 370 + 220 = 590.

An estimate does not give an accurate answer, but it will be close.

Think of all the numbers that will round to 260.
Think of all the numbers that will round to 370.
What do you notice about these numbers?

Guided practice

Estimate the answer of each calculation by rounding the numbers to the nearest 10.

a 232 + 458

$$230 + 460 = 690$$

b 568 − 324

$$570 - 320 = 250$$

Lesson 4: **Rounding to the nearest 100**

- Round 3-digit numbers to the nearest 100

Key words
- **hundreds**
- **tens**
- **ones**
- **position**
- **estimate**
- **round**
- **round up**
- **round down**

Number

Let's learn

Remember!

round **down**   0 10 20 30 40

50 60 70 80 90   round **up**

To work out 569 + 323, estimate the answer first.

Rounding is a useful strategy to use for **estimating**.

We know we can estimate to the nearest 10, now we will round to the nearest 100.

569 rounds **up** to 600.

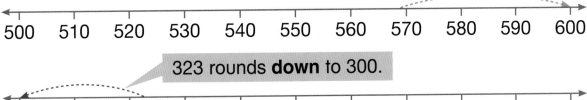

500   510   520   530   540   550   560   570   580   590   600

323 rounds **down** to 300.

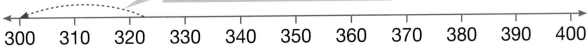

300   310   320   330   340   350   360   370   380   390   400

An estimate of 569 + 323 is 600 + 300 = 900.

Rounding to the nearest 100 will give an estimate but it is not as close as rounding to the nearest 10.

There are 100 numbers that round to 100.
They start at 50 and go up to 149!

How many numbers will round to 200?

What is the smallest number that rounds to 200? What is the largest?

How many numbers will round to 600?

What is the smallest number that rounds to 600? What is the largest?

Guided practice

Mark 326 on the number line.

Show whether you would round it **up** or **down** to the nearest 100.

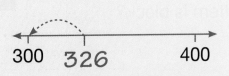

300   *326*   400

Lesson 1: **Equal parts: quantities**

- Understand and explain that fractions are one or more equal parts and all the parts, taken together, equal one whole

Key words
- **fraction**
- **unit fraction**
- **non-unit fraction**
- **whole**
- **part**
- **equal**
- **denominator**
- **numerator**

Let's learn

These part–whole diagrams show different fractions.

One whole
Two halves $\dfrac{1}{2}$

numerator

denominator

We write two thirds as $\dfrac{2}{3}$.

One whole
Three thirds $\dfrac{1}{3}$

One whole
Four quarters (or fourths) $\dfrac{1}{4}$

We write three quarters as $\dfrac{3}{4}$.

One whole
Five fifths $\dfrac{1}{5}$

We write four fifths as $\dfrac{4}{5}$.

If the whole orange bar is worth 25, what is $\dfrac{1}{5}$?
What is $\dfrac{2}{5}$? What about $\dfrac{3}{5}$? What about $\dfrac{4}{5}$?

Guided practice
What fraction of each tile pattern is black?

a 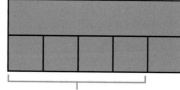 $\dfrac{1}{4}$

b 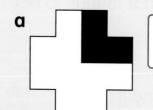 $\dfrac{3}{5}$

Lesson 2: **Equal parts: shapes**

* Understand that fractions are several equal parts of a shape

Key words
* **fraction**
* **unit fraction**
* **non-unit fraction**
* **whole**
* **part**
* **equal**
* **denominator**
* **numerator**

Number

Let's learn

Fractions of shapes are based on how much space they take up, inside the whole.

These two rectangles have both been divided into quarters but in different ways. Each whole rectangle has been halved and then halved again.

Each part is $\frac{1}{4}$ of the whole.

Even though the parts are different shapes, they all take up the same amount of space. Each triangle is half of a square, so takes up the same amount of space as a small rectangle.

What fraction of this shape is shaded?

What fraction is not shaded?

How many parts make up the whole shape?

Guided practice

What fraction of the shape is shaded?

a

$\frac{1}{5}$ of this shape is shaded

b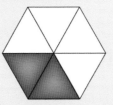

$\frac{2}{5}$ of this shape is shaded

c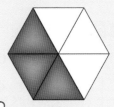

$\frac{3}{5}$ of this shape is shaded

Number

Lesson 3: **Same fraction, different whole**

Key words
- **fraction**
- **unit fraction**
- **non-unit fraction**
- **whole**
- **part**
- **equal**
- **denominator**
- **numerator**

- Understand the relationship between 'whole' and 'parts'

Let's learn

These shapes all have one quarter shaded.

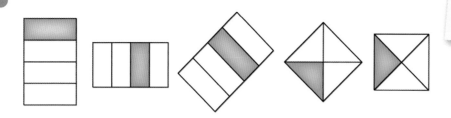

It doesn't matter that they are different shapes and different sizes, or how they are positioned.

In each shape, 1 of the 4 equal parts making the whole is shaded.

These shapes do **not** show quarters.

1 of 4 parts of each is shaded, but in each shape the parts are **not** equal.

Which of these is the odd one out?

Talk to your partner about your reasons.

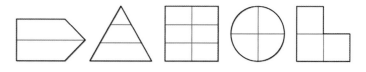

Guided practice

Explain why each of these shapes shows $\frac{1}{4}$.

Each shape has 4 equal parts. One part in each has been shaded and this is $\frac{1}{4}$

Lesson 4: **Fractions of quantities and sets**

Key words
* fraction
* whole
* part
* equal
* denominator
* numerator

Number

* Understand that fractions describe equal parts of a quantity or set of objects

Let's learn

These 7 cherries are $\frac{1}{2}$
of the cherries in a bag.
So there must be 7 cherries in the other half.
There are 14 cherries in the bag.

7	

14	
7	7

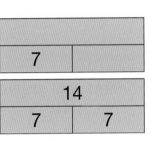

These 5 biscuits are $\frac{1}{4}$
of the amount in a packet.
If we know this, we know that there must be
$4 \times \frac{1}{4}$ in the whole packet.
There are 20 biscuits in the whole packet.

5			

20			
5	5	5	5

Talk to your partner about this part–whole model.
What are the missing parts?
What fraction is each part?
What is the whole?

6			

Guided practice

Leona has 25 marbles. She gives $\frac{1}{5}$ of them to her friend.
How many does she give to her friend?
Draw a part–whole model.

25				
5	5	5	5	5

Lesson 1: **Same value, different appearance**

- Recognise that two fractions can have the same value

Key words
- **fraction**
- **whole**
- **part**
- **equal**
- **denominator**
- **numerator**
- **equivalent**

Let's learn

1			
$\frac{1}{2}$		$\frac{1}{2}$	
$\frac{1}{4}$	$\frac{1}{4}$	$\frac{1}{4}$	$\frac{1}{4}$

whole

halves

quarters

The fraction wall shows:

$$\frac{2}{2} = 1 \qquad \frac{4}{4} = 1 \qquad \frac{1}{2} = \frac{2}{4}$$

1				
$\frac{1}{5}$	$\frac{1}{5}$	$\frac{1}{5}$	$\frac{1}{5}$	$\frac{1}{5}$
$\frac{1}{10}$ $\frac{1}{10}$	$\frac{1}{10}$ $\frac{1}{10}$	$\frac{1}{10}$ $\frac{1}{10}$	$\frac{1}{10}$ $\frac{1}{10}$	$\frac{1}{10}$ $\frac{1}{10}$

whole

fifths

tenths

This fraction wall shows that:

$$\frac{5}{5} = 1 \qquad \frac{10}{10} = 1 \qquad \frac{1}{5} = \frac{2}{10}$$

 Look at the second fraction wall above.

What fraction is the same as $\frac{5}{10}$?

Guided practice

Colour the second square to show the same fraction as shown in the first one. Write the new fraction.

$$\frac{2}{8}$$

Lesson 2: **Equivalent fractions**

- Recognise that two fractions can have the same value

Key words
- **fraction**
- **whole**
- **part**
- **equal**
- **denominator**
- **numerator**
- **equivalent**

Let's learn

Equivalent fractions are fractions that have the same value.

They have different numerators and denominators.

Remember!

$$\frac{1}{2}$$ — numerator — denominator

This fraction wall shows that $\frac{2}{2} = 1$ $\frac{10}{10} = 1$ $\frac{1}{2} = \frac{5}{10}$.

This number line shows other fractions that are equivalent to $\frac{1}{2}$.

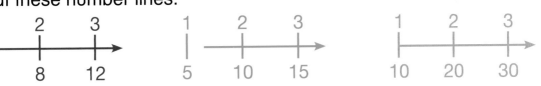

Look at these number lines.

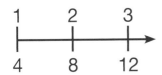 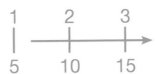

Write some other fractions that are equivalent to $\frac{1}{4}$, $\frac{1}{5}$ and $\frac{1}{10}$.

Look at the numerators and denominators. What patterns do you notice?

Guided practice

Draw a ring around the shapes that are shaded to show $\frac{1}{2}$.

71

Lesson 3: **Comparing fractions**

Number

- Compare two fractions

Key words
- fraction
- whole
- part
- equal
- denominator
- numerator
- larger than
- smaller than
- compare

Let's learn

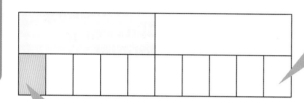

The tenths are smaller than the halves.

This fraction wall shows that one half is larger than one tenth of the same whole.

$$\frac{1}{2} > \frac{1}{10} \quad \text{or} \quad \frac{1}{10} < \frac{1}{2}$$

$\frac{3}{10}$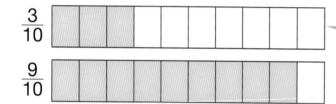

$\frac{9}{10}$

These bars have both been divided into tenths. Both fractions have the same denominator.

To compare fractions with the same denominators we look at the numerators: $9 > 3$

There are more parts in $\frac{9}{10}$ than in $\frac{3}{10}$.

$$\frac{9}{10} > \frac{3}{10} \quad \text{or} \quad \frac{3}{10} < \frac{9}{10}$$

Which of these is the largest fraction? $\frac{1}{8} \quad \frac{1}{4} \quad \frac{1}{10} \quad \frac{1}{12} \quad \frac{1}{6}$
Which is the smallest?
Talk to your partner about how you know.

Guided practice

Write >, < or = between each pair of fractions.

a $\frac{1}{5}$ $\boxed{<}$ $\frac{1}{4}$ **b** $\frac{1}{4}$ $\boxed{<}$ $\frac{1}{2}$ **c** $\frac{2}{4}$ $\boxed{=}$ $\frac{5}{10}$ **d** $\frac{1}{8}$ $\boxed{>}$ $\frac{1}{10}$

Lesson 4: **Ordering fractions**

• Order a set of fractions

Key words
• **fraction**
• **whole**
• **part**
• **equal**
• **denominator**
• **numerator**
• **largest**
• **smallest**
• **order**

Number

Let's learn

If we can **compare** fractions, we can **order** them too.

Remember! The unit fraction with the largest denominator is the smallest.

1							
$\frac{1}{2}$				$\frac{1}{2}$			
$\frac{1}{4}$		$\frac{1}{4}$		$\frac{1}{4}$		$\frac{1}{4}$	
$\frac{1}{8}$	$\frac{1}{8}$	$\frac{1}{8}$	$\frac{1}{8}$	$\frac{1}{8}$	$\frac{1}{8}$	$\frac{1}{8}$	$\frac{1}{8}$

In this fraction wall $\frac{1}{8}$ is the smallest fraction and $\frac{1}{2}$ is the largest.

From smallest to largest, the order of these fractions is: $\frac{1}{8}$ $\frac{1}{4}$ $\frac{1}{2}$

From largest to smallest, the order of these fractions is: $\frac{1}{2}$ $\frac{1}{4}$ $\frac{1}{8}$

We can order the fractions on a number line:

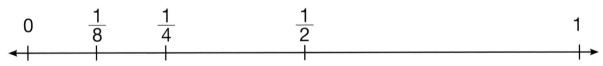

Talk to your partner about these fractions.
What fractions are they?

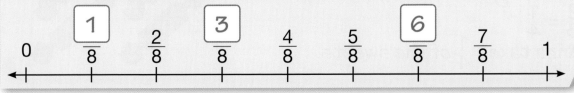

Order them from largest to smallest.

Guided practice

The fractions are in order. Write the missing numbers.

0 $\boxed{1}\over 8$ $\frac{2}{8}$ $\boxed{3}\over 8$ $\frac{4}{8}$ $\frac{5}{8}$ $\boxed{6}\over 8$ $\frac{7}{8}$ 1

Number

Lesson 1: **Fractions and division (A)**

- Understand that a fraction can be represented as a division

Let's learn

Fractions are a way of **sharing**.
We can think of **division** as sharing.

Remember!

$$\frac{1}{2}$$

numerator

denominator

The numerator shows how many parts we are considering.

The denominator shows how many equal parts the whole is divided into.

$\frac{1}{2}$ is $1 \div 2$

$\frac{1}{4}$ is $1 \div 4$

$\frac{3}{4}$ is $3 \div 4$

Make up a number story to explain how $\frac{1}{2}$ is the same as $1 \div 2$.

Guided practice

There are 8 sweets. Anna takes 2 of them.
What fraction of the sweets does she take?

She takes 2 out of 8.

This is the same as $2 \div 8$

$2 \div 8 = \frac{2}{8}$

$\frac{2}{8} = \frac{1}{4}$

Anna takes $\frac{1}{4}$ of the sweets.

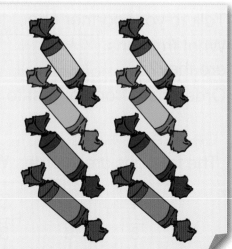

Number

Lesson 2: **Fractions and division (B)**

- Recognise fractions as operators

Let's learn

To find a fraction of an amount, divide by the denominator and multiply by the numerator.

To find $\frac{1}{4}$ of 28, first divide 28 by 4.

$28 \div 4 = 7$

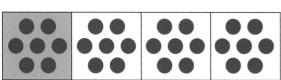 So, $\frac{1}{4}$ of 28 = 7

To find $\frac{3}{4}$ of 28, first divide 28 by 4 to find $\frac{1}{4}$.

$28 \div 4 = 7$

Then multiply the answer by 3.

$7 \times 3 = 21$

For a unit fraction such as $\frac{1}{4}$ there is no need to multiply.

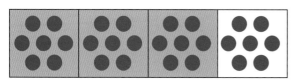 So, $\frac{3}{4}$ of 28 = 21

Spot the mistake!

Talk to your partner about which fraction statement is wrong and why.

$\frac{1}{4}$ of 24 = 6 $\frac{1}{2}$ of 24 = 12 $\frac{3}{4}$ of 24 = 18

$\frac{1}{3}$ of 24 = 7 $\frac{1}{10}$ of 50 = 5

Guided practice

Write the fraction statement that matches each division calculation. You do not need to work out the answers.

a $6 \div 3 = \boxed{\frac{1}{3} \text{ of } 6}$ **b** $16 \div 4 = \boxed{\frac{1}{4} \text{ of } 16}$

Lesson 3: **Adding and subtracting fractions (A)**

- Add and subtract fractions with the same denominator

Key words
- fraction
- whole
- part
- equal
- denominator
- numerator
- add
- subtract
- commutative
- inverse

Let's learn

Adding and subtracting fractions is like adding and subtracting whole numbers.

The denominators of the fractions are the same because they tell you the number of parts in the whole.

Add or subtract the numerators to find out how many parts of the whole you have.

$$\frac{5}{8} + \frac{2}{8} = \frac{7}{8}$$

Remember! Addition is commutative. So, $\frac{2}{8} + \frac{5}{8} = \frac{7}{8}$

Remember! The inverse relationship between addition and subtraction.

So, $\frac{7}{8} - \frac{2}{8} = \frac{5}{8}$ and $\frac{7}{8} - \frac{5}{8} = \frac{2}{8}$

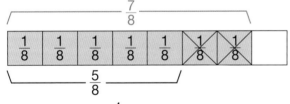

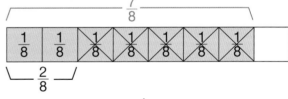

Suzie reads $\frac{1}{4}$ of her book before dinner. She reads $\frac{1}{4}$ after dinner. What fraction of her book does she still have to read?

Talk to your partner about how you know.

Guided practice

Luke says: I can add $\frac{1}{4}$ and $\frac{1}{4}$. My answer is $\frac{2}{8}$.

Do you agree with Luke?   No

Explain why. The denominators are the same. He should have just added the numerators. The answer should be $\frac{2}{4}$ or $\frac{1}{2}$.

Number

Lesson 4: **Adding and subtracting fractions (B)**

- Add and subtract fractions with the same denominator

Key words
- fraction
- whole
- part
- equal
- denominator
- numerator
- add
- subtract
- commutative
- inverse

Let's learn

When we add or subtract fractions with the same denominator, the denominator stays the same.

That's because it is the number of parts in the whole.

Just add or subtract the numerators.

$$\frac{3}{10} + \frac{4}{10} = \frac{7}{10}$$

Remember!
Addition is commutative.

$$\frac{4}{10} + \frac{3}{10} = \frac{7}{10}$$

Remember! The inverse relationship between addition and subtraction.

$$\frac{7}{10} - \frac{3}{10} = \frac{4}{10} \text{ and } \frac{7}{10} - \frac{4}{10} = \frac{3}{10}$$

Tom eats $\frac{2}{5}$ of his fruit in the morning.

He eats $\frac{1}{5}$ of his fruit in the afternoon.

What fraction of his fruit does he still have to eat?

Talk to your partner about how you know.

Guided practice

Paulo says:

Do you agree? | Yes |

If I cut my cake into five parts and eat three of them, I will have $\frac{2}{5}$ left.

Explain why. You can draw a diagram.

$\frac{5}{5} - \frac{3}{5} = \frac{2}{5}$

Lesson 1: **Units of time**

• Choose suitable units to measure time

Key words
• time
• second
• minute
• hour
• day
• week
• month
• year

Let's learn

We use different units of time to measure the length of different activities.

We would measure something we did very quickly in **seconds**.

Sneezing takes seconds!

A football match lasts for 90 **minutes**.

It might take a few **minutes** to read a few pages of a book.

It will take **hours** to read a whole book.

We sleep for about 8 **hours** each night.
Days are much longer than **minutes**.
We might go on holiday for two **weeks**.
Age is measured in **years**.

Make a list of things that take hours to do.
Explain to your partner why these things take hours.

Guided practice
What can you do in about one hour? Write three different things.

<u>Watch a tv show. Cook a meal. Wash and clean a car</u>

Geometry and Measure

Lesson 2: **Telling the time (A)**

> • Read and write the time to five minutes

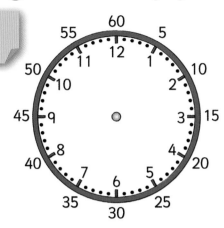

Let's learn

We can count in 5s to tell 'minutes past times'. We can link these times to digital times.

quarter past 10

10 fifteen

10:15

40 minutes past 5

20 minutes to 6

5 forty

05:40

Geometry and Measure

Spot the mistake!

Which digital time is wrong? Talk to your partner about why it is wrong.

10:25   8:50   11:5

4:10   11:55   6:40

Guided practice

Write the digital time to match each analogue clock.

a

b

06:25

03:50

Lesson 3: **Telling the time (B)**

• Read and write the time

Key words
• time
• minute
• hour
• analogue
• digital
• minutes past
• minutes to

Let's learn

We can tell the time to the nearest 5 minutes on analogue and digital clocks.

Can we tell the time to the nearest minute?

21 minutes past 9

9 twenty-one

09:21

49 minutes past 7

7 forty-nine

07:49

11 minutes to 8

This clock has lost its hour hand. What time could it be?

How many times can you think of?

Talk to your partner about this.

Guided practice

Write the digital time to match each analogue clock.

a
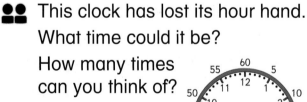

b

08:13   **10:43**

Lesson 4: **Timetables**

- Use and interpret timetables
- Find time intervals

Leave Ayetown	Arrive Beeville
9:15	10:50
10:00	11:45
11:20	12:30
12:00	1:15
1:20	2:40
2:05	3:30
3:15	4:40
4:00	4:45
4:20	5:25
5:35	6:20
5:50	6:55

Let's learn

A timetable is a plan of times at which events happen. These are usually shown in hours and minutes.

This timetable shows the time the train leaves Ayetown and the time it arrives at Beeville.

We can see from the timetable that one of the trains leaves Ayetown at 3:35 and arrives at Beeville at 4:45.

Find the time that the 3rd train arrives in Beeville. What time did it leave Ayetown?

Write the departure and arrival times on paper.

Do this again for another train and then another.

Geometry and Measure

Guided practice

What time did the 2:05 train from Ayetown arrive in Beeville?

3:30

Lesson 1: **2D shapes**

- Identify, describe, name and sketch regular and irregular 2D shapes

Geometry and Measure

Let's learn

Polygons are 2D shapes with three or more straight sides.

triangle   square   pentagon   hexagon

heptagon   octagon   nonagon

These polygons are **regular**, their sides are all equal and they have angles of equal size.

These polygons are **irregular**.

quadrilateral   pentagon   hexagon   octagon   decagon

Quadrilaterals are 2D shapes with four straight sides.

Can you make four identical triangles from this square?
Do you think that the triangles will be regular or irregular?
Talk to your partner about how you can prove what you think.

Guided practice

Draw a ring around the regular polygons.

Lesson 2: **Sorting 2D shapes**

- Sort regular and irregular 2D shapes

Let's learn

Remember!

- **Regular** shapes have sides that are the same length and angles of equal size.
- **Polygons** are 2D shapes with three or more straight sides.

Non-polygons are shapes that have at least one curved side.

Here are some examples of non-polygons.

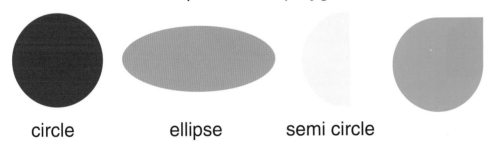

circle   ellipse   semi circle

We can sort these shapes into polygons and non-polygons.

Talk to your partner about another way to sort the shapes.

How many different ways can you think of to sort the shapes?

Guided practice

Write in the headings to show how these shapes have been sorted.

Regular shapes	Not regular shapes

Geometry and Measure

Lesson 3: **Symmetry**

- Identify horizontal and vertical lines of symmetry on 2D shapes

Let's learn

A shape or pattern is symmetrical if we can divide it into two identical halves that mirror each other. The straight line is called a **line of symmetry**.

These shapes have a **vertical** line of symmetry.

These shapes have a **horizontal** line of symmetry.

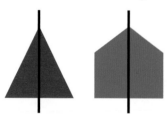

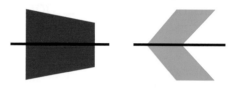

These shapes have both **vertical** and **horizontal** lines of symmetry.

These shapes have **no** lines of symmetry.

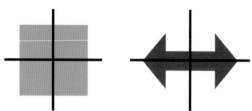

Draw a shape with a horizontal line of symmetry on a piece of paper. How can you prove the shape is symmetrical?

Draw another shape that has both vertical and horizontal lines of symmetry.

Guided practice
Draw one line of symmetry on each of these shapes.

Lesson 4: **Angles**

- Compare angles with a right angle
- Recognise that a straight line is equal to two right angles or a half turn

Key words
- **right angle**
- **quarter turn**
- **half turn**

Geometry and Measure

Let's learn

One right angle is the same as one quarter of a turn.

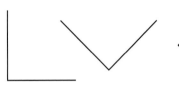 quarter turn clockwise

quarter turn anticlockwise

There are right angles everywhere!
Squares and rectangles have four right angles.

These right angles
are in different
positions.

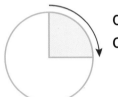

If we put two right angles together we get a straight line.
Two right angles = Half a turn.

Which of these angles are greater than a right angle?
Which are less than a right angle?

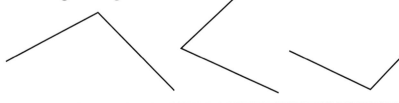

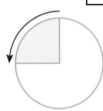

Guided practice
Draw a ring around the right angles in these shapes.

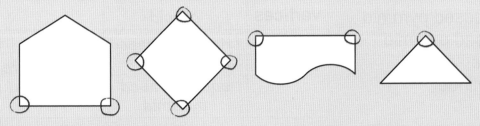

Lesson 1: **Identifying 3D shapes**

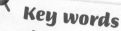

Key words
- **edge**
- **face**
- **surface**
- **flat**
- **curved**
- **vertex**
- **vertices**

- Identify, describe, classify, name and sketch 3D shapes
- Recognise pictures, drawings and diagrams of 3D shapes

Let's learn

These 3D shapes have flat faces, straight edges and vertices.

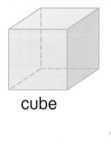

cube

cuboid

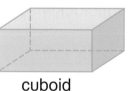

square-based
pyramid

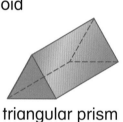

triangular prism

triangular-based
pyramid or tetrahedron

The most common 3D shapes with flat faces are prisms and pyramids.

Some 3D shapes
have at least
one curved surface.

sphere

hemisphere

cylinder

If you put two identical cubes together, face on face, what other shape can you make?

Talk to your partner about the properties of the new shape.

Can you see anything that is the same shape in the classroom?

Guided practice

Write the names of the shapes on the Carroll diagram:
triangular-based pyramid
square-based pyramid
hemisphere   cube
sphere   cylinder
cuboid

	all faces are flat	all faces are not flat
8 vertices	cube cuboid	
not 8 vertices	triangular-based pyramid square-based pyramid	sphere hemisphere cylinder

Lesson 2: **Prisms**

- Identify, describe, classify, name and sketch prisms
- Recognise pictures, drawings and diagrams of prisms

Key words
- **prism**
- **polygonal**
- **edge**
- **face**
- **vertex**
- **vertices**

Geometry and Measure

Let's learn

Prisms have two faces in the shape of polygons at the ends.
All the other faces are rectangles.

| triangular prism | cube | cuboid | pentagonal prism | hexagonal prism | octagonal prism |

The number of rectangular faces is the same as the number of sides of the polygon.

A triangular prism has two triangular faces and three rectangular faces.

This table shows how many faces, edges and vertices some prisms have.

Shape	Faces	Edges	Vertices
triangular prism	5	9	6
cube or cuboid	6	12	8
pentagonal prism	7	15	10
hexagonal prism	8	18	12
octagonal prism	10	24	16

Choose a prism, but don't tell your partner which one you are thinking of.

Describe the properties of your prism to your partner.

Can they tell you what your prism is?

Swap and play again.

Guided practice
Draw a ring around all the prisms.

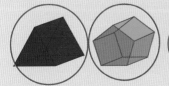   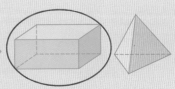

Lesson 3: **Pyramids**

- Identify, describe, classify, name and sketch pyramids
- Recognise pictures, drawings and diagrams of pyramids

Key words
- **pyramid**
- **edge**
- **face**
- **vertex**
- **vertices**

Geometry and Measure

Let's learn

A pyramid is a 3D shape that has a polygon as its base.
This gives the pyramid its name. All the other faces are triangles.

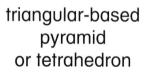

triangular-based pyramid or tetrahedron

square-based pyramid

pentagonal-based pyramid

hexagonal-based pyramid

This table shows the properties of the pyramids above.

Pyramid	Faces	Edges	Vertices
triangular	4	6	4
square	5	8	5
pentagonal	6	10	6
hexagonal	7	12	7

 Find two identical square-based pyramids.

Place them base on base.

Sketch the shape you have made.

Label its properties.

Tell your partner its properties.

Guided practice
Draw a ring around all the pyramids.

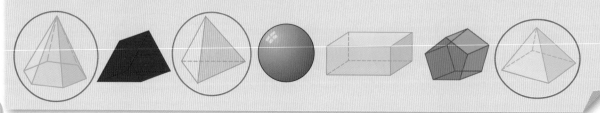

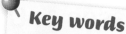
Lesson 4: **3D shapes in real life**

- Identify, describe, classify, name and sketch 3D shapes
- Recognise pictures, drawings and diagrams of 3D shapes

Key words
- prism
- pyramid
- edge
- face
- surface
- flat
- curved
- vertex
- vertices

Let's learn

Now we can identify and describe shapes, we should be able to recognise them in real life.

What 3D shapes do you recognise?

Look at this shape man.

What shapes can you see? Identify as many as you can with your partner.

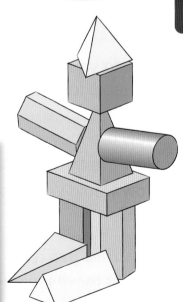

Guided practice

Draw a ring around the object that is the odd one out.

Explain why.

The battery is a cylinder and all the other objects are prisms.

Geometry and Measure

Lesson 1: **Units of length**

- Estimate and measure lengths in centimetres (cm), metres (m) and kilometres (km)
- Understand the relationship between units

Let's learn

We measure short lengths, like a book, with a ruler.

We measure long lengths and heights with a metre stick.

Key words
- centimetres (cm)
- metres (m)
- kilometres (km)
- ruler
- metre stick
- tape measure
- trundle wheel
- length
- height
- width
- distance

We use a tape measure to measure longer lengths and widths.

We use a trundle wheel to measure the length of a path.

Facts
1 metre = 100 centimetres
1 kilometre = 1000 metres

These facts lead to new facts.

Talk to your partner about the tools for measuring length.

How are they alike?

How are they different?

How would you use each one?

5 m = 500 cm   200 cm = 2 m

1 m = 100 cm

7 m = 700 cm   900 cm = 9 m

Guided practice
Order these lengths, from shortest to longest.
37 m 12 km 45 cm 12 m 5 km 28 cm

28 cm, 45 cm, 12 m, 37 m, 5 km, 12 km

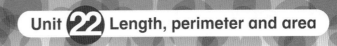

Lesson 2: **Measuring lines**

- Estimate and measure lengths in centimetres (cm)
- Use instruments that measure length

Key words
- centimetre (cm)
- ruler
- estimate
- actual

Let's learn

Use a ruler to measure the length of a line. Make sure the end of the line is at the 0 mark on the ruler.

The length of this line is 15 cm.

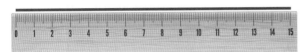

The width of this pencil sharpener is 1 cm.

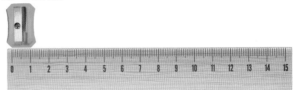

The length of this fork is 18 cm.

It is always helpful to estimate length before measuring.

Geometry and Measure

Jamie says:

I measured my line. It is 22 cm long.

Talk to your partner about what Jamie has done wrong.

What is the real length of his line?

Guided practice

Holly draws a line. It is 8 cm long.

a. Use your ruler to draw a line half the length of Holly's line.

Half of 8 cm is 4 cm.

b. Use your ruler to draw a line twice the length of Holly's line.

Double 8 cm is 16 cm.

Lesson 3: **Perimeter**

Key words
- perimeter
- distance
- centimetre (cm)

- Understand that perimeter is the total distance around a 2D shape and can be calculated by adding lengths

Geometry and Measure

Let's learn

The length of this tennis court is 24 m.

Its width is 11 m.

The distance all the way round the tennis court is:

24 m + 11 m + 24 m + 11 m, which is 70 m.

The distance around the outside of a shape is called the **perimeter**.

The perimeter of the tennis court is 70 m.

We should always estimate perimeters before working them out. An estimate for the perimeter of the tennis court might be 60 m (20 + 20 + 10 + 10).

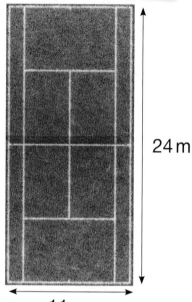

24 m

11 m

A quick way to work out the perimeter of a rectangle is to add the length and width and double it.

24 + 11 = 35
35 × 2 = 70

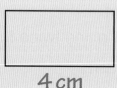 This is a square. Each side is 6 cm.

Discuss how you could work out the perimeter of this square.

What is its perimeter?

Guided practice

Draw a rectangle that is 4 cm long and 2 cm wide.

Work out its perimeter.

2 cm

4 cm

4 cm + 2 cm
= 6 cm × 2 =
12 cm

Lesson 4: **Area**

• Understand that area is how much space a 2D shape occupies within its perimeter

Let's learn

The perimeter of this baseball court is the distance around the outside.

It is the total length of all of its four sides.

The **area** of the baseball court is the amount of space inside its perimeter.

We can measure an area in **square units**.

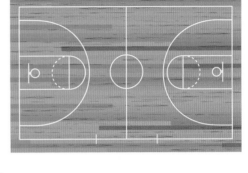

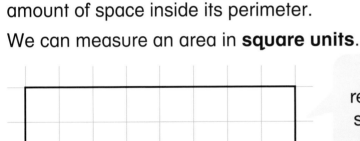

To find the area of this rectangle, we can count the squares inside it. There are 24 squares so the area is 24 square units.

There are 8 squares in each row. There are 3 rows.
8 × 3 = 24
The area is 24 squares.

Look at the square on page 92. Each side is 6 cm.

Discuss how you could work out the area of the square.

What is its area?

Geometry and Measure

Guided practice

Sophie is building a patio.

She uses 15 square paving slabs.

She decides that she will lay 3 rows of 5 slabs.

a. Draw a diagram of Sophie's patio.

b. What is the area of her patio? *15 square units*

Lesson 1: **Units of mass**

- Estimate and measure mass in kilograms and grams
- Understand the relationship between units

Geometry and Measure

Let's learn

Mass is how heavy something is.

We measure mass in kilograms and grams.

We use **grams** to find the mass of items that are quite **light**.

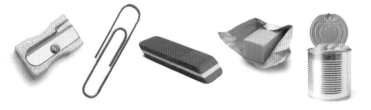

We use **kilograms** to find the mass of items that are quite **heavy**.

Fact
1 kilogram = 1000 grams

This fact leads to new facts.

5 kg = 5000 g

2000 g = 2 kg

1 kg = 1000 g

7 kg = 7000 g

9000 g = 9 kg

If you know that 1000 g = 1 kg, discuss with your partner what other facts you know. Can you think of some that involve fractions of a kilogram?

Guided practice

Order these masses, from heaviest to lightest.

$\frac{1}{4}$ kg   600 g   2 kg   150 g   40 kg   55 g

40 kg, 2 kg, 600 g, $\frac{1}{4}$ kg, 150 g, 55 g

Lesson 2: **Measuring in kilograms**

- Estimate and measure mass in kilograms
- Use instruments that measure mass

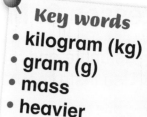

Let's learn

Remember! The mass of light objects is measured in grams. The mass of heavier things is measured in kilograms.

This person is looking at their mass.
The unit of measure isn't given.
He could be 60 g or 60 kg.
Which do you think it is?

This bag of rice has a mass of 1 kg.

Each of these dumb bells has a mass of 1 kg.

Look at the mass of the watermelon.

Talk to your partner about how you know its mass is closer to 3 kg than 2 kg.

Can you give a closer estimate of the mass of the watermelon?

Guided practice

a Read the mass on the scale.

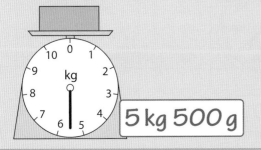

5 kg 500 g

b Draw the pointer on the scale to show the mass.

3 kg

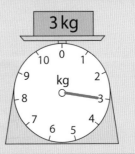

Geometry and Measure

Lesson 3: **Measuring in grams**

- Estimate and measure mass in grams
- Use instruments that measure mass

Geometry and Measure

Let's learn

Remember! Heavier items are measured in kilograms. Lighter items are measured in grams.

The pointer on a scale may be on, or near, a whole number of kilograms or a division between kilograms.

On this scale, the orange's mass is less than 1 kg, so its mass is measured in grams.

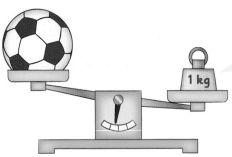

On this balance scale the pan with the 1 kg weight is lower than the pan with the ball.

This means that the ball is lighter than 1 kg and therefore is measured in grams.

What mass is shown on the scale?

Talk to your partner about the different weights on the scale. What could they be?

Guided practice

a Read the mass on the scale.

400 g

b Draw the pointer on the scale to show the mass.

750 g

Lesson 4: **Measuring in kilograms and grams**

- Estimate and measure mass in kilograms and grams
- Understand the relationship between units
- Use instruments that measure mass

Key words
- kilogram (kg)
- gram (g)
- mass
- heavier
- lighter

Geometry and Measure

Let's learn

Masses may be measured in kilograms or grams.

Often, mass is measured using both kilograms and grams.

This scale shows the mass of 1600 grams.

Remember! 1 *kilogram* = 1000 *grams*

To convert 1600 g to kilograms and grams:

$1600\,g = 1000\,g + 600\,g$
So, the mass is 1 kg and 600 g.
Write it as 1 kg 600 g.

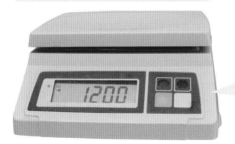

This scale shows 1200 g.
1200g is the same as 1kg 200g

Just like any numbers, we can add and subtract mass.

With your partner, make up two masses in kilograms and grams.
Find the total mass. Find the difference between the two.

Guided practice
The pan has a total mass of 1 kg 750 g.
The empty pan has a mass of 200 g.

What is the mass of the stew? | 1 kg 550 g |
Show how you worked out your answer.

$1\,kg\,750\,g - 200\,g \quad 750 - 200 = 550 \quad 1\,kg\,550\,g$

Lesson 1: **Units of capacity**

- Estimate and measure capacity in litres and millilitres
- Understand the relationship between units

Key words
- **capacity**
- **litre (l)**
- **millilitre (ml)**
- **equivalent**

Let's learn

We measure the capacity of small containers in **millilitres**.
We measure the capacity of large containers in **litres**.

Fact
1 litre = 1000 millilitres

$5l = 5000\,ml$

$9000\,ml = 9l$

This fact leads to new facts.

$1l = 1000\,ml$

$7l = 7000\,ml$

$2000\,ml = 2l$

👥 Look at these containers.
Would you use litres or millilitres to measure their capacities?
Are there any you are not sure about?
Tell your partner why you are unsure.

Guided practice

Order these capacities, from smallest to largest.

7l	600 ml	$\frac{1}{4}l$	300 ml	$\frac{1}{2}l$	1l

$\frac{1}{4}l$, 300 ml, $\frac{1}{2}l$, 600 ml, 1l, 7l

Geometry and Measure

98

Lesson 2: **Measuring capacity**

- Estimate and measure capacity in litres and millilitres
- Understand the relationship between units

Let's learn

Capacity is the amount of liquid a container can hold when it is full.

We measure capacities of bottles in litres.

2 litres

This bottle holds 2 litres.
Its capacity is 2 litres.

The capacity of a bucket is measured in litres.

We use millilitres to measure the capacity of a teaspoon.

To find the capacity of a container, we can use measuring jugs or measuring cylinders.

Look carefully at these measuring jugs and cylinders.

What is the same about them?

What is different?

Discuss this with your partner.

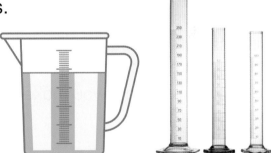

Guided practice

Maisie has 1 *l* 500 m*l* of juice in a jug. She pours 700 m*l* into a bottle.

How much is left in the jug? | 800 *ml* |
Show your working.

| 1 *l* 500 ml = 1500 ml   1500 ml – 700 ml = 800 ml |

Geometry and Measure

Geometry and Measure

Lesson 3: **Measuring in litres and millilitres**

 Key words
- **capacity**
- **litre (*l*)**
- **millilitre (m*l*)**
- **equivalent**
- **measuring cylinder**

- Estimate and measure capacity in litres and millilitres
- Understand the relationship between units
- Use instruments that measure capacity

Let's learn

We can use a measuring jug or measuring cylinder to measure the capacity of a container.

Most jugs and cylinders show a scale, similar to a number line. We read the scale to find out how much liquid is in the container.

Reading the scale on this measuring cylinder, we can see the amount of liquid is 750 m*l*.

Look at these three measuring cylinders. Discuss with your partner what's the same and what's different about them.

What statements can you make comparing cylinders A, B and C?

A   B   C

Guided practice

Mark measured the capacity of a tub. He noticed that the capacity was between 1*l* 500 m*l* and 1*l* 700 m*l*.

Give three possible capacities.

1*l* 550 m*l*, 1*l* 600 m*l*, 1*l* 650 m*l*

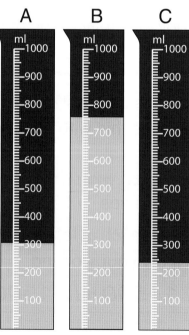

Lesson 4: **Temperature**

• Use instruments that measure temperature

Let's learn

We measure temperature in **degrees**.

We usually use the Celsius scale. The unit is °C.

Some countries use the Fahrenheit scale.
The unit is °F.

We use thermometers to measure temperature.

These are some of the different thermometers
that we could use.

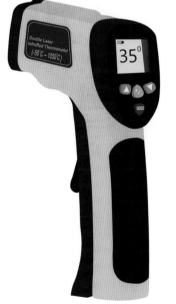

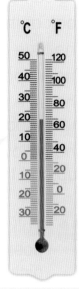

We would use
this thermometer
to measure the
temperature in
a room.
It says that the
temperature is
21°C or 70°F.

Geometry and Measure

Guided practice
Shade each
thermometer to
show the correct
temperature.

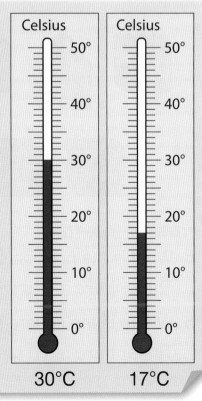

Celsius — 30°C

Celsius — 17°C

Look at the three
thermometers above.
What statements can
you and your partner
make comparing the
temperatures they show?

Lesson 1: **Position**

• Use language associated with position

Let's learn

The position of someone or something is the place where it is.

We can use lots of words to describe position.

Look at this cupboard.

The blue cup is **in front of** the green plate.

The teapot is **beside** the green jug.

The blue plate is **inside** the cupboard.

The yellow cup is **between** the blue cup and the green cup.

Talk to your partner about the position of the sphere in these pictures.

Key words
• position
• on top
• above
• below
• beside
• between
• underneath
• inside

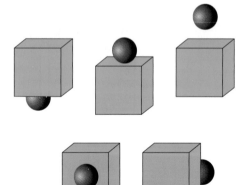

Guided practice

a What is the position of the sphere?

The sphere is between the two cubes.

b Where is the sphere now?

The sphere is to the right of the cube.

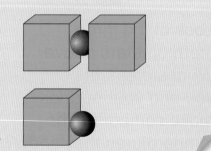

Lesson 2: **Direction and movement**

- Use language associated with direction and movement

Let's learn

To travel from one place to another, we move in different directions.

Here are some of the directions we travel in.

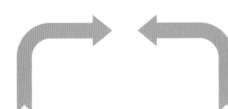

| right | left | straight | clockwise | anticlockwise |

Remember!

Anticlockwise is the opposite direction to the movement of the hands on an analogue clock.

Clockwise is the same direction as the movement of the hands on an analogue clock.

Give your partner different directions to get from one part of the classroom to another.

Use the **Key words** in the box at the top of the page.

Geometry and Measure

Guided practice

If we make a one quarter turn clockwise and then another, what size turn have we made?

We have made a half turn.

Where will we be facing if we make a whole clockwise turn?

We will be facing the same way as we started

Lesson 3: **Compass points**

- Use the four compass points to describe position, direction and movement

Let's learn

We can use a compass to find directions.

This compass is pointing **south**.
If we travel in the opposite direction, we will be going **north**.

This compass is pointing **west**.
If we travel in the opposite direction, we will be going **east**.

Patsy is facing south. She thinks that to get from the start to the finish, she must go:
- south for 3 squares
- east for 3 squares
- south for 3 squares
- west for 4 squares
- south for 2 squares

Do you agree? Why?

Talk to your partner about it.

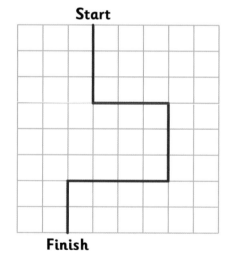

Guided practice

You are facing north. What direction will you face if you make a quarter turn anticlockwise?

I will *face west.*

You are facing west. What direction will you face if you make a three-quarter turn anticlockwise?

I will *face north.*

Lesson 4: **Reflections**

• Sketch reflections of 2D shapes

Key words
• **reflect**
• **horizontal**
• **vertical**
• **mirror line**
• **line of symmetry**

Let's learn

What do you see when you look at yourself in a mirror?

You see your **reflection**.

Your reflection looks just like you, but the sides of your face are swapped across a vertical mirror line.

A reflection is an image of anything as it would be seen in a mirror.

You will often see **vertical** or **horizontal** reflections.

This shows the reflection of a shape across a vertical mirror line.

This shows the reflection of a shape across a horizontal mirror line.

A **mirror line** is another way of describing a **line of symmetry**.

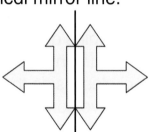

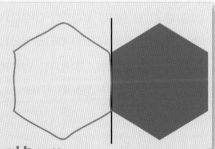

Work with a partner.

Sketch this shape on a piece of paper.

Draw a vertical mirror line like this.

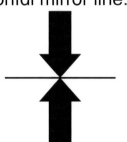

One of you reflect the shape across this mirror line.

Next draw a horizontal mirror line under the new shape and reflect this.

Now draw a mirror line on the left of your new shape and reflect it again.

Talk to each other about what you notice.

Guided practice

Here is a hexagon. One of its edges is along a vertical mirror line.

Draw its reflection.

What do you notice?

My reflection has made a symmetrical pattern.

Geometry and Measure

Lesson 1: **Venn diagrams**

- Record, organise, represent and interpret data in Venn diagrams

Let's learn

There are lots of ways to sort things.

One way is to use a **Venn diagram**.

We use rules to sort things into the Venn diagram.

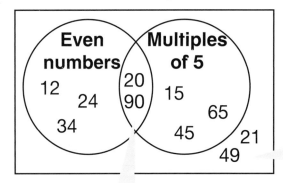

The two numbers outside are neither even nor multiples of 5. This is called a universal set. It contains all the numbers we are interested in. The circles inside the rectangle have sorted the numbers into even numbers and multiples of 5.

If anything matches both the rules, we place it in the part where the sets intersect.

Draw a two criteria Venn diagram.

6 Talk to your partner about how you could sort these numbers:

1, 5, 6, 8, 12, 15, 17, 18, 24, 27

Make sure to label your Venn diagram.

Guided practice

Here are two Venn diagrams, each made from one set.

If we combine them, which numbers will go in the intersection?

12 and 24

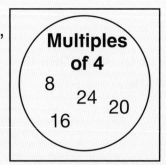

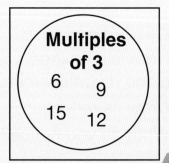

Statistics and Probability

Lesson 2: **Carroll diagrams**

- Record, organise, represent and interpret data in Carroll diagrams

Let's learn

In a Carroll diagram we sort according to whether something obeys a rule or does not obey that rule.

Even numbers	Not even numbers
10 38 20 24	15 21 49 65

We can use a Carroll diagram to sort according to two rules.

In this Carroll diagram the same numbers are sorted.

	Even numbers	Not even numbers
Multiples of 5	10 20	15 65
Not multiples of 5	38 24	49 21

What is the difference between the two Carroll diagrams?
What is the same?

Look at the second Carroll diagram above. Talk to your partner about what it shows.

Write down four statements that are true.

Guided practice

Look at this Carroll diagram.

	Multiples of 2	Not multiples of 2
Multiples of 5	10 20	25 15
Not multiples of 5	26 32	23 49

Which numbers are **not** multiples of 2 or 5? $\boxed{23, 49}$

Which numbers are multiples of both 2 and 5? $\boxed{10, 20}$

Statistics and Probability

107

Lesson 3: **The statistical cycle**

Key words
* **organise**
* **represent**
* **data**
* **rule**
* **criteria**
* **tally chart**

• Recording, organising, representing and interpreting data

Let's learn

Remember! We can record, organise and represent data in different ways, including tally charts, block graphs and pictograms.

Statistical cycle

This cycle can help us to work through the different stages of a statistical investigation or to answer a question to a problem that needs solving.

1. What's the problem? Make a plan

4. Discuss the data and check predictions

2. Record, organise and represent the data

3. Interpret the data

What data should we collect to solve the problem?

How should we collect the data?

How should we represent it?

Discuss this tally chart with your partner.
What statements and conclusions can you make?

Number of people who went into a shop	
Bakery	ᴜᴜᴜ ᴜᴜᴜ IIII
Supermarket	ᴜᴜᴜ III
Post Office	ᴜᴜᴜ ᴜᴜᴜ ᴜᴜᴜ ᴜᴜᴜ IIII
Newsagent	ᴜᴜᴜ ᴜᴜᴜ ᴜᴜᴜ II
Shoe shop	ᴜᴜᴜ ᴜᴜᴜ IIII

Guided practice

Write three statements about the tally chart.

There are 20 learners in Class B.

There are more learners in Class D than in Class C.

Class A has the fewest learners.

Learners in Stage 3	
Class	**Tally**
Class A	ᴜᴜᴜ III
Class B	ᴜᴜᴜ ᴜᴜᴜ ᴜᴜᴜ ᴜᴜᴜ
Class C	ᴜᴜᴜ ᴜᴜᴜ II
Class D	ᴜᴜᴜ ᴜᴜᴜ ᴜᴜᴜ

Statistics and Probability

Lesson 4: **Frequency tables**

- Record, organise, represent and interpret data in frequency tables

Let's learn

All the children in a class were asked to choose their favourite subject. This table shows the results.

The tally column shows the results as tallies.

The frequency column shows the results as numbers.

Subject	Tally	Frequency
Maths	卌 卌 l	11
Reading	卌 ll	7
Science	卌	5
Art	卌 卌	10
Music	卌 l	6

 Draw the tally chart above but only include the Subject column. Then ask other learners about their favourite subject and write the results in the Tally column. Then, fill in the frequency column.

Guided practice

The frequency column in this table has been completed. Fill in the tally column.

Hobby	Tally	Frequency
Reading	卌 卌 卌	15
Films	卌 卌 卌 llll	19
Drawing	卌 卌 卌 卌 ll	22
Sports	卌 卌 lll	13

Statistics and Probability

Lesson 1: **Pictograms**

- Record, organise, represent and interpret data in pictograms

Key words
- record
- organise
- represent
- data
- criteria
- pictogram

Let's learn

Tally charts and frequency tables are two ways to represent data.

A **pictogram** is another way of representing data using pictures or symbols.

This pictogram shows how many goals the football teams in the Planet league scored last season.

Football team	Goals
Mars	●●●●
Jupiter	●●●●●●
Saturn	●●●●●●●◖
Pluto	●◖
Mercury	●●●●●
Venus	●●●●●●●◖

● = 2 goals

👥 Look at the Planet league pictogram and take turns to ask each other questions about the data.

Guided practice
Use the pictogram to answer these questions.

a How many goals did Mars score? Mars scored 8 goals.

b How many more goals did Saturn score than Pluto?
Saturn scored 14 more goals than Pluto.

c How many goals did Mercury and Venus score altogether?
Mercury and Venus scored 25 goals altogether.

Statistics and Probability

Lesson 2: **Bar charts**

• Record, organise, represent and interpret data in bar charts

Key words
• record
• organise
• represent
• data
• criteria
• bar chart
• vertical axis
• interval

Let's learn

A **bar chart** is another way to represent data.

Bar charts are similar to pictograms, but they use bars instead of symbols.

A bar chart has a vertical axis marked in steps.

James' family are sorting out their shoes.

The bar chart shows how many pairs of shoes they each have.

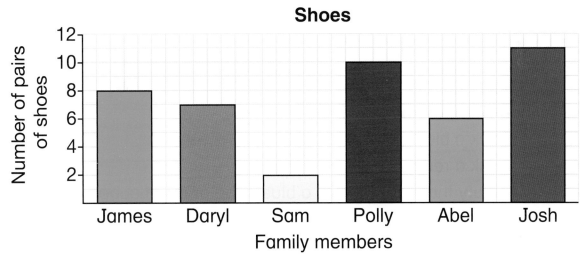

Shoes — Number of pairs of shoes vs Family members (James 8, Daryl 7, Sam 2, Polly 10, Abel 6, Josh 11)

👥 Look at the Shoes bar chart and take turns to ask each other questions about the data.

Guided practice
Use the bar chart to answer these questions.

a How many more pairs of shoes does Josh have than Daryl?
Josh has 4 more pairs.

b How many pairs of shoes do Polly and Abel have altogether?
Polly and Abel have 16 pairs of shoes altogether.

c How many pairs of shoes does James' family have altogether?
They have 44 pairs of shoes altogether.

Statistics and Probability

111

Lesson 3: **Chance (1)**

- Use familiar language associated with chance to describe events

Key words
- chance
- will happen
- might happen
- will not happen

Let's learn

Chance is the possibility of something happening.

Some things definitely **will** happen.

> Tomorrow will come after today.

Some things definitely **will not** happen.

> You will grow an extra head.

Some things **might** happen.

> It will rain next month.

Imagine you put these cubes in a bag, shake the bag and pick out one cube, without looking.

- Might you pick a blue cube?
- Might you pick a red cube?
- Is it **more likely** that you will pick a blue cube or a red cube?

Look at this statement:

> I will see someone I know on the way home from school.

What is the chance of this happening?

Talk about this with your partner and explain your thinking.

Guided practice

How likely are these things to happen?

Choose one of these phrases: will happen   might happen
                                               will not happen

a April will come after March.   <u>will happen</u>

b It will snow in the mountains in winter.   <u>might happen</u>

c I will turn five on my next birthday.   <u>will not happen</u>

Statistics and Probability

112

Lesson 4: **Chance (2)**

- Conduct chance experiments, and present and describe the results

Let's learn

Think about tossing this coin.
It may land heads up or it may land tails up.

Tails   Heads

Landing heads up **might happen** and landing tails up **might happen**.

There is the **same chance** of getting either heads or tails.

It is difficult to say whether tossing the coins would show heads or tails.

Here are three digit cards.   **4** **4** **4**

If we turn them face down and pick one, we **will** pick a 4.

Now look at these three digit cards.   **4** **7** **4**

If we turn them face down and pick one, we are **more likely** to pick a 4 than a 7.

But we could pick a 7. We can say that both **might happen**.

You place the digit cards 2, 2, 2, 5 and 9 face down on the table in front of you and pick one.

Which number are you most likely to pick?

Why?

Will you pick a 2 more often than you pick a 5 or 9? Why is that?

Guided practice

If you put these cubes in a bag and pick one, you are most likely to pick a red.

Do you agree? Yes

Why? Because there are more red cubes than green.

The Thinking and Working Mathematically Star

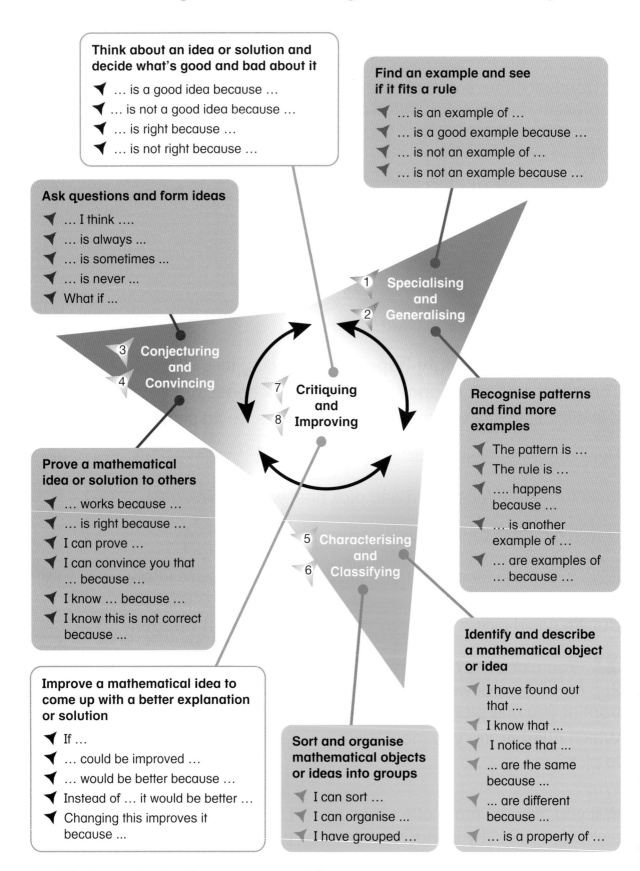

Think about an idea or solution and decide what's good and bad about it
- … is a good idea because …
- … is not a good idea because …
- … is right because …
- … is not right because …

Find an example and see if it fits a rule
- … is an example of …
- … is a good example because …
- … is not an example of …
- … is not an example because …

Ask questions and form ideas
- … I think ….
- … is always …
- … is sometimes …
- … is never …
- What if …

1 2 Specialising and Generalising

3 4 Conjecturing and Convincing

7 8 Critiquing and Improving

Recognise patterns and find more examples
- The pattern is …
- The rule is …
- …. happens because …
- … is another example of …
- … are examples of … because …

Prove a mathematical idea or solution to others
- … works because …
- … is right because …
- I can prove …
- I can convince you that … because …
- I know … because …
- I know this is not correct because …

5 6 Characterising and Classifying

Identify and describe a mathematical object or idea
- I have found out that …
- I know that …
- I notice that …
- … are the same because …
- … are different because …
- … is a property of …

Improve a mathematical idea to come up with a better explanation or solution
- If …
- … could be improved …
- … would be better because …
- Instead of … it would be better …
- Changing this improves it because …

Sort and organise mathematical objects or ideas into groups
- I can sort …
- I can organise …
- I have grouped …

Acknowledgements

Photo acknowledgements

Every effort has been made to trace copyright holders. Any omission will be rectified at the first opportunity.

p9t Kentoh/Shutterstock; p10b Neirfy/Shutterstock; p25t Pineapple studio/Shutterstock; p25c Tim UR/Shutterstock; p25b Alexander Mak/Shutterstock; p55tl Olga Popova/Shutterstock; p55tr Ukki Studio/Shutterstock; p25bl Hlemeida Ivan/Shutterstock; p25br Maxstockphoto/Shutterstock; p56tl Brent Hofacker/Shutterstock; p56tr Louella938/Shutterstock; p56cl Jacek Chabraszewski/Shutterstock; p56cr Annata78/Shutterstock; p56b Africa Studio/Shutterstock; p57t BlueRingMedia/Shutterstock; p57b Aratehortua/Shutterstock; p69t Anna Kucherova/Shutterstock; p69b Nattika/Shutterstock; p76b Normallens/Shutterstock; p77b Baibaz/Shutterstock; p78t VectorsMarket/Shutterstock; p78bl Igor Krylytsya/Shutterstock; p78br Bergamont/Shutterstock; p79 Szefei/Shutterstock; p81 Oleksiy Mark/Shutterstock; p83 Icon99/Shutterstock; p89tl Nikola Bilic/Shutterstock; p89tcl Sofiaworld/Shutterstock; p89tcr Jemastock/Shutterstock; p89tr Kabardins photo/Shutterstock; p89cl MIKHAIL GRACHIKOV/Shutterstock; p89ccl Aleksandrs Bondars/Shutterstock; p89ccr Irin-k/Shutterstock; p89cr Africa Studio/Shutterstock; p89bl Ojal/Shutterstock; p89bcl DaniiD/Shutterstock; p89bcr Stuar/Shutterstock; p89br Digital Assets/Shutterstock; p90tl Avector/Shutterstock; p90tr Evikka/Shutterstock; p90bl Seregam/Shutterstock; p90br Weerastudio/Shutterstock; p91tl Lunewind/Shutterstock; p91tr Mega Pixel/Shutterstock; p91c Simo988/Shutterstock; p91b Grebeshkovmaxim/Shutterstock; p92 Zooropa/Shutterstock; p93 Anucha Tiemsom/Shutterstock; p94tl Mega Pixel/Shutterstock; p94tcl Magicoven/Shutterstock; p94tc Ajt/Shutterstock; p94tcr Bonchan/Shutterstock; p94tr New Africa/Shutterstock; p94bl Gts/Shutterstock; p94bcl Brovko Serhii/Shutterstock; p94bcr Topseller/Shutterstock; p94br Den Rozhnovsky/Shutterstock; p95t Vladislav Gashchuk/Shutterstock; p95cl Walnut Bird/Shutterstock; p95cr Mtphoto19/Shutterstock; p95b Morphart Creation/Shutterstock; p97t Vereschagin Dmitry/Shutterstock; p97b Denio109/Shutterstock; p98t Cigdem/Shutterstock; p98bl Bessarab/Shutterstock; p98bcl Andrey Eremin/Shutterstock; p98bc DenisMArt/Shutterstock; p98bcr Andrey_Kuzmin/Shutterstock; p98br Susan Schmitz/Shutterstock; p99t HitToon/Shutterstock; p99cl Onair/Shutterstock; p99cr Andrey Eremin/Shutterstock; p99bl AlexanderZam/Shutterstock; p99br PRILL/Shutterstock; p101tl Igor Semenov/Shutterstock; p101tr Tomas Ragina/Shutterstock; p101b Mawa Art/Shutterstock; p102t Ivonne Wierink/Shutterstock; p103t Giraphics/Shutterstock; p103b Colin Cramm/Shutterstock; p105t Tivota Irina/Shutterstock; p105b WEB-DESIGN/Shutterstock.